Wind Energy

Solar Energy R&D in the European Community

Series G

Wind Energy

Volume 2

Publication arrangements: T. C. Jones

Solar Energy R&D
in the European Community

Series G Volume 2

Wind Energy

Potential of Wind Energy in the European Community
An Assessment Study

by

H. SELZER
ERLO - Raumfachtechnik GmbH, Bremen, F.R.G.

D. Reidel Publishing Company
A MEMBER OF THE KLUWER ACADEMIC PUBLISHERS GROUP
Dordrecht / Boston / Lancaster / Tokyo

for the Commission of the European Communities

Library of Congress Cataloging in Publication Data
Main entry under title:

Selzer, H.
Potential of wind energy in the European community.

(Solar energy R & D in the European Community.
Series G, Wind energy ; v. 2)
Bibliography: p.
1. Wind power–European Economic Community countries.
I. Commission of the European Communities. II. Title. III. Series.
TJ820.S45 1986 333.79'2 86–456
ISBN 90-277-2205-6

CIP

Publication arrangements by
Commission of the European Communities
Directorate-General Information Market and Innovation, Luxembourg

EUR 10376

Published by D. Reidel Publishing Company
P.O. Box 17, 3300 AA Dordrecht, Holland

Sold and distributed in the U.S.A. and Canada
by Kluwer Academic Publishers,
190 Old Derby Street, Hingham, MA 02043, U.S.A.

In all other countries, sold and distributed
by Kluwer Academic Publishers Group,
P.O. Box 322, 3300 AH Dordrecht, Holland

Printed in The Netherlands

CONTENTS

PREFACE

The present book provides a technical and macro-economical assessment of the wind energy potential in Europe. It is an important result of the first Wind Energy R&D Programme of the European Communities which was started in 1980.

The potential electricity production of wind energy conversion systems in the European Community has been analysed in a concerted action with an international group of experts under the leadership of Dr. Selzer from the company ERNO/MBB. In order to develop reliable data, it was necessary to analyse the wind which is available in the various parts of Europe, study in detail all candidate sites for the possible installation of wind turbines, and to review the technological cost-effectiveness of wind turbines. The availability of utility networks for grid-connection of the wind turbines was also included in the analysis.

Three important facts are described for the first time in this book: First, all member countries of the European Community are concerned by wind energy utilization. There is a high wind energy potential in Northern as well as in Southern Europe. Secondly, most of the potential areas for wind energy utilization are along the coastal zones. Thirdly, electricity contribution from wind energy converters to Europe's overall consumption could eventually be very large. In principle, it would even be possible to provide all of Europe's electricity needs from wind energy.

However, in the foreseeable future, the actual development of this technology in Europe will be much more modest because at the present time, wind turbines are not yet fully cost-competitive with the conventional means of electricity production.

In any case, this book clearly shows that the promise of wind energy in Europe is such that its development deserves full attention and it has become urgent to set up an aggressive strategy for implementation.

I take this opportunity to thank all the participants of the study for their cooperation and I am confident that this book will be an important help to inform a wider public of this very interesting and new technology.

W. Palz

W. PALZ
Head of the Wind Energy R&D Programme

1. INTRODUCTION

Within the framework of its Solar Energy R&D programme, the Commission of the European Communities has initiated a programme for the investigation on a European level of problems related to wind energy. The study started in November 1981 and was subdivided into five different topics:

Action 1: Assessment of the technical/economic prospects for wind energy in the EC

Action 2: Wind energy data

Action 3: R&D related topics of grid interface problems

Action 4: Compilation of available data of existing wind generators

Action 5: Support to the collection of new data on existing wind generators.

In order to report as accurately as possible on the national situations of the 10 countries involved, experts from the research institutes or the industries of each country were contracted to report to the Action coordinators. The task of the coordinators was to formulate a coordinated approach, stimulate the discussion to reveal problems of interest, gather the national findings together, and present the combined results in a balanced form.

The contractors of the Actions contributed by reporting on the various national regulations for, and results in the field of wind energy, and they also recommended topics of further interest for the team.

This volume presents the results of Actions 1 and 3 which were performed separately by the same coordinator. Although an attempt was made to avoid the expression of individual opinions, the contributions of the national contractors should not be taken as national standpoints on the matters in discussion. The same is true for the coordinator. Though most results were approved jointly by the teams, there is no doubt about the personal influence of the chairman in the preparation of meetings and work instructions and in the presentation and interpretation of the results.

One difficulty encountered in the study was that of finding the best compromise between European conformity and national specialities, particularly as great differences in the involvement in this new technology existed at the time of writing.

Another aspect influencing the presentation of the findings concerns the preliminaries. Many of the data collected may be deemed rough estimates, or stem from pilot plants, and corrections and improvements must be expected.

Every effort was made to derive firm statements, while still giving the reader the possibility to repeat the assessment by using different individual assumptions or updated sources.

It had been a pleasure to cooperate with the experts of the countries involved and the coordinator expresses his gratitude to the participants of Action 1 and 3 for their excellent contributions and stimulating discussions during the course of the project.

Participants of the study:

Action 1: Assessment of the Technical/Economic Prospects for Wind Energy in the European Countries

Coordinator	Mr. H. Selzer	MBB-ERNO	Bremen
Belgium	Mr. Ch. Hirsch	Vrije Universiteit	Brussels
Denmark	Mr. J. Fenhann	Risö National Lab.	Roskilde
France	Mr. A. Haentjens	SEMA	Montrouge
West-Germany	Mr. G. Obermair	ATW-GmbH	Regensburg
Greece	Mr. N. Athanassiadis	Nat. Techn. Univ.	Athens
Ireland	Mr. J. White	IIRS	Dublin
Italy	Mr. E. Sesto	ENEL	Milan
Netherlands	Mr. A. Janssen	ECN	Petten
United Kingdom	Mr. M. Mytton	W.S. Atkins Planning	Epsom

Action 3: R&D Related Topics of Grid Interface Problems

Coordinator	Mr. H. Selzer	MBB-ERNO	Bremen
Belgium	Mr. Ch. Hirsch	Vrije Universiteit	Brussels
Denmark	Mr. P. Nielsen	DEFU	Lyngby
France	Mr. A. Gourod	Aerowatt	Paris
West-Germany	Mr. G. Obermair	ATW-GmbH	Regensburg
Greece	Mr. G. Vachtsevanos	University of Thrace	Xanthi
Ireland	Mr. H. Power	University College	Dublin
Italy	Mr. A. Invernizzi	ENEL/CREL	Milan
Netherlands	Mr. N. Halberg	KEMA	Arnhem
United Kingdom	Mr. G. Gardner	CEGB	London

Fig. 2-0 Europe photographed by satellite in June 1982
with indicated isovents

2. ASSESSMENT OF THE CONVERTIBLE POTENTIAL OF WIND ENERGY IN COUNTRIES OF THE EUROPEAN COMMUNITIES

Many estimates of wind energy on national and global bases have been published since Betz's fundamental report on wind turbine flow in 1926. The call for new evaluations is becoming more frequent as a result of the continuous improvement of technology and the data now available. Yet land is not so readily available owing to an intensified awareness of the importance of using our land to serve the best economic purposes and at the same time protecting the natural environment. Is the public likely to accept visual pollution of the landscape as the price of reduced chemical emmissions? For that reason, the method selected has been based on certain well defined restrictions in order to derive figures controllable by everyone. Restrictions stemming from subjective opinions have also been taken into consideration, as will be seen in the following discussion, but the final answer has been left to the reader.

In the method applied, the areas of interest were first subdivided into subfields, using lines of equal wind speeds (isovents) and geographical longitude and latitude as boundaries. A mesh of triangular grid points was superposed on a detailed map, and the ratio of the total grid points to those acceptable as potential sites for large wind energy converters (WECs) was taken.

The number of potential sites for small WECs was obtained by adding up the number of remote dwellings within each subfield or by taking these figures from statistical sources.

It has been assumed that small WECs will be operated by private owners, and that large-scale electricity production will be undertaken by the Electricity Boards using large WECs. The question of clusters of small WECs has been omitted because these need very detailed siting analysis, and local wind mapping is beyond the scope of this study. During the course of this study it was confirmed that large-scale wind energy penetration will be on the basis of large WECs and not small WECs.

Special attention has been given to providing the calculations and results in a form which will allow for updating if new and more sophisticated data become available.

2.1 Distribution of Mean Annual Wind Speeds

The degree of applicability of wind data for the calculation of wind energy potential differs very much from country to country because measurements and evaluations have not always been carried out for this specific purpose. Therefore, within the scope of the first European wind energy programme, Action 2 has been defined to prepare what is called briefly a "Wind Atlas".

As the long time needed for the preparation of the Wind Atlas precluded its use in the assessment study, a solution on a preliminary basis was arrived at by admitting the existing data for the time being; the results will be updated when improved data become available. The problem arises from the standardized measurement of wind speed at 10 m height at the synoptic stations, thus producing data which are strongly influenced by surface roughness. Extrapolation from this level to the 100 m layer may be very erroneous. A simplified picture of the European wind zones is presented in the illustration in fig. 2-0.

A more reliable approach has been developed in Denmark [2-1] by starting from the "free stream" at altitudes of more than 1000 m and calculating the wind profile from its dependence on friction losses due to surface roughness. As this method cannot yet be applied to countries where the terrain differs from that of Denmark, the assessment study had to use the isovent data for the 10 m level for the sake of comparability.

In order to facilitate the interpretation of the energy potential and economics of the following chapters, a brief description of some wind characteristics is given here.

In most cases, the annual distribution of wind speeds follows a "Weibull" distribution

$$f(v) = a \cdot c \cdot v^{c-1} \cdot \exp(-a \cdot v^c)$$

where c is the shape parameter and $A = a^{-1/c}$ is a scaling parameter.

As illustrated in Fig. 2-1, the Weibull function describes a great variety of distributions. For c = 1 the distribution is exponential, c = 2 represents the Rayleigh distribution often used in wind energy, and c = 3.5 gives a Gaussian distribution.

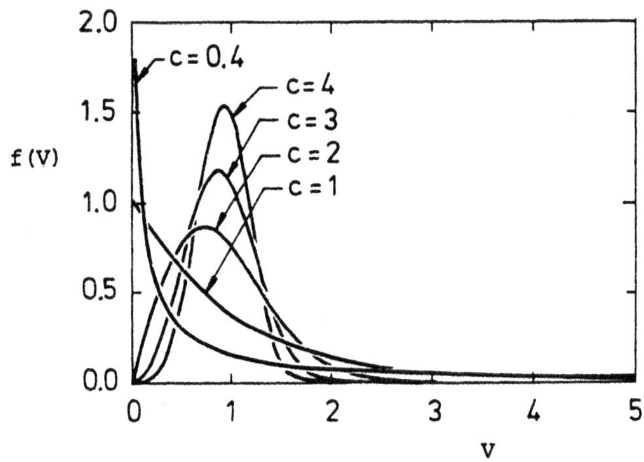

Fig. 2-1: The importance of the shape parameter c for the shape of the Weibull distribution

Accumulation of f (v), e.g. integration, leads to

$$f(v) = 1 - \exp(-a \cdot v^c)$$

representing the probability of a wind speed equal to or less than v. Wind data in this form are often presented as an "annual duration curve".

The mean annual wind speed $\bar{v}$ may be derived from:

$$\bar{v} = A \cdot \Gamma\left(1 + \frac{1}{c}\right)$$

where Γ represents the gamma function.

The assessment study is based on the measured mean annual wind speeds without control of the shape parameter (due to non-availability), and this will cause some minor shifts of the results when improved wind measurements become available.

According to boundary flow theories, the wind profile above ground may be described by an exponential increase with height, hence the ratio of wind speed at different heights - under normal conditions - can be calculated by the Hellmann equation:

$$\frac{v_{h_1}}{v_{h_2}} = \left(\frac{h_1}{h_2}\right)^{\alpha}$$

where h stands for the different heights.

The exponent α represents the influence of surface roughness and thermal layers on the wind shear, ranging between 0.15 and 0.40. A value of $\alpha = 1/7$ has been found to be widely applicable to low surface roughness.

As this power law postulates ideal conditions, it is evident that extrapolation of the wind speeds measured at 10 m height to 100 m height may lead to considerable errors. This should be reduced in the forthcoming evaluation of the Action 2 "Wind Atlas".

As mentioned above, a more consistent approach resulted from the Danish Wind Programme which started from the undisturbed geostrophic wind at 1000 m height. In plotting the speed decrease on a logarithmic scale, a lowest height value is found at zero wind speed, called roughness length. The complete systematic of how to use this model for the calculation of local wind energy is described in the Wind Atlas for Denmark [2-1]. The classification of the countryside is listed in Table 2-1. There are four different roughness classes depending on the structure of the terrain. Sometimes, these classes are also addressed as wind zones.

Although the Danish treatment cannot be applied to other countries without careful consideration of all boundary conditions, this system has been very helpful in the course of the study as it allows a systematic ordering of the otherwise complex data structure for the existing wind turbines in different locations.

Roughness Class	Terrain	Roughness Length	Relative Energy
0	water areas	0 - 1 mm	10
1	open country areas with very few bushes, trees, and buildings	1 cm	7
2	farmland with scattered buildings and hedges with separation in excess of 1000 m	5 cm	6
3	built-up areas, forests, and farmland with many hedges	30 cm	3

Table 2-1:
Types of Terrain, Roughness Classes, and Roughness Lengths

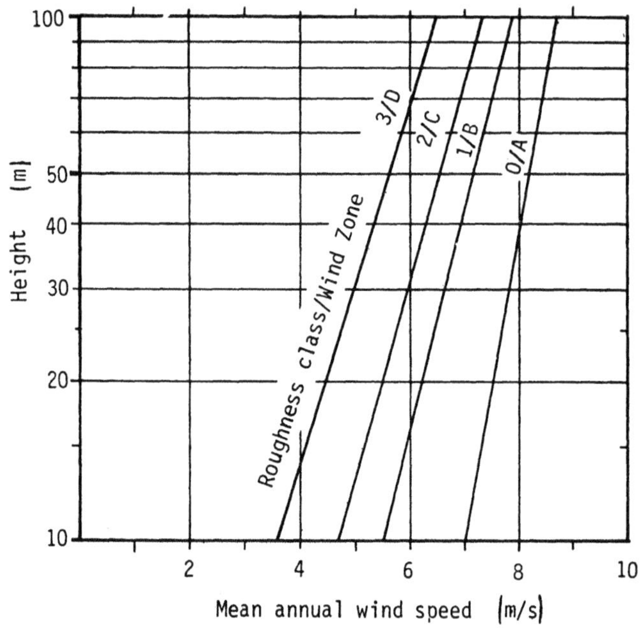

Fig. 2-2: Wind Speed Dependence on Height

As a framework for the references, the wind speed dependence on height in the four different classes is plotted in Fig. 2-2 and may serve as a first approach until more specific local data become available.

The wind data supplied by the national contractors for this study are presented in Figures 2-3 to 2-11, and more national details are given in a special CEC-report.

Since annual mean wind speeds from 142 stations and the local surface roughness lengths around them are available, they have been taken into account for the evaluation of mean wind speed at 100 m height in the selected fields. Coastal regions with wind speeds greater than 4 m/s are indicated by dotted lines in Fig. 2-11.

The wind speeds at 100 m height were calculated by using the Hellmann power equation.

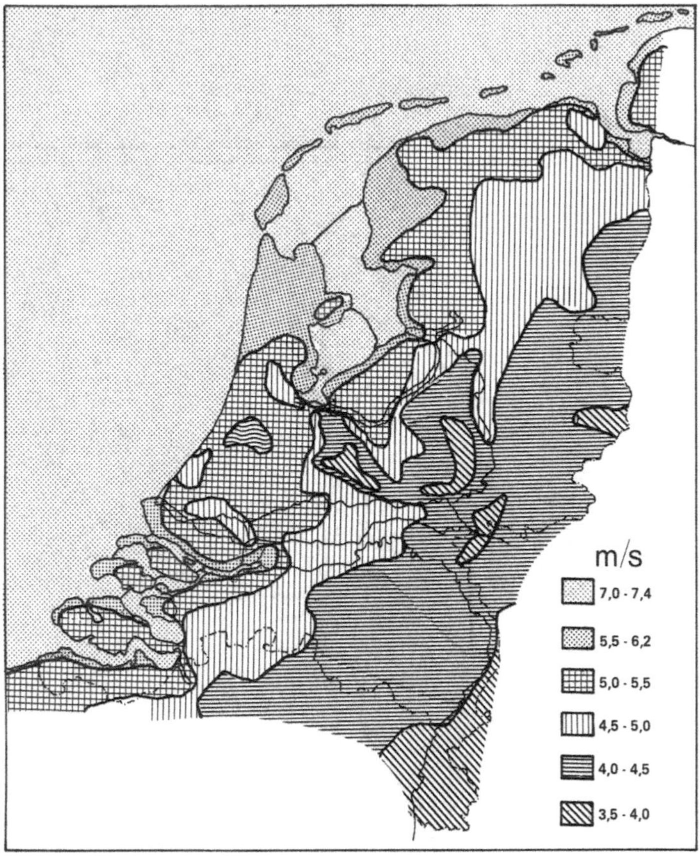

Fig. 2-3: 10 m Isovent Map of NETHERLANDS

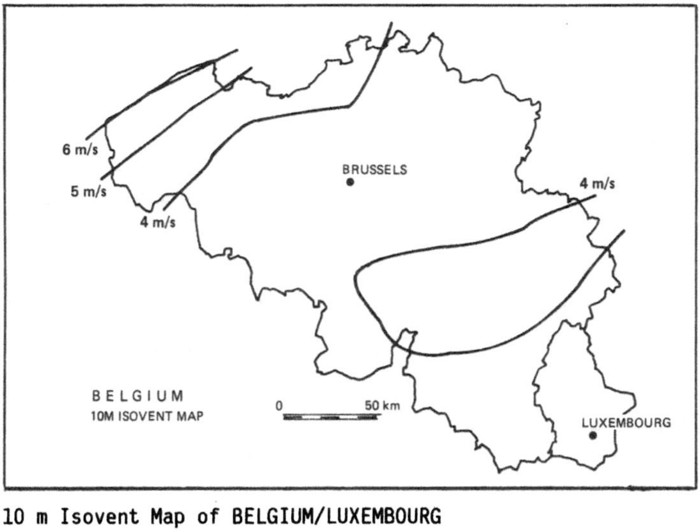

Fig. 2-4: 10 m Isovent Map of BELGIUM/LUXEMBOURG

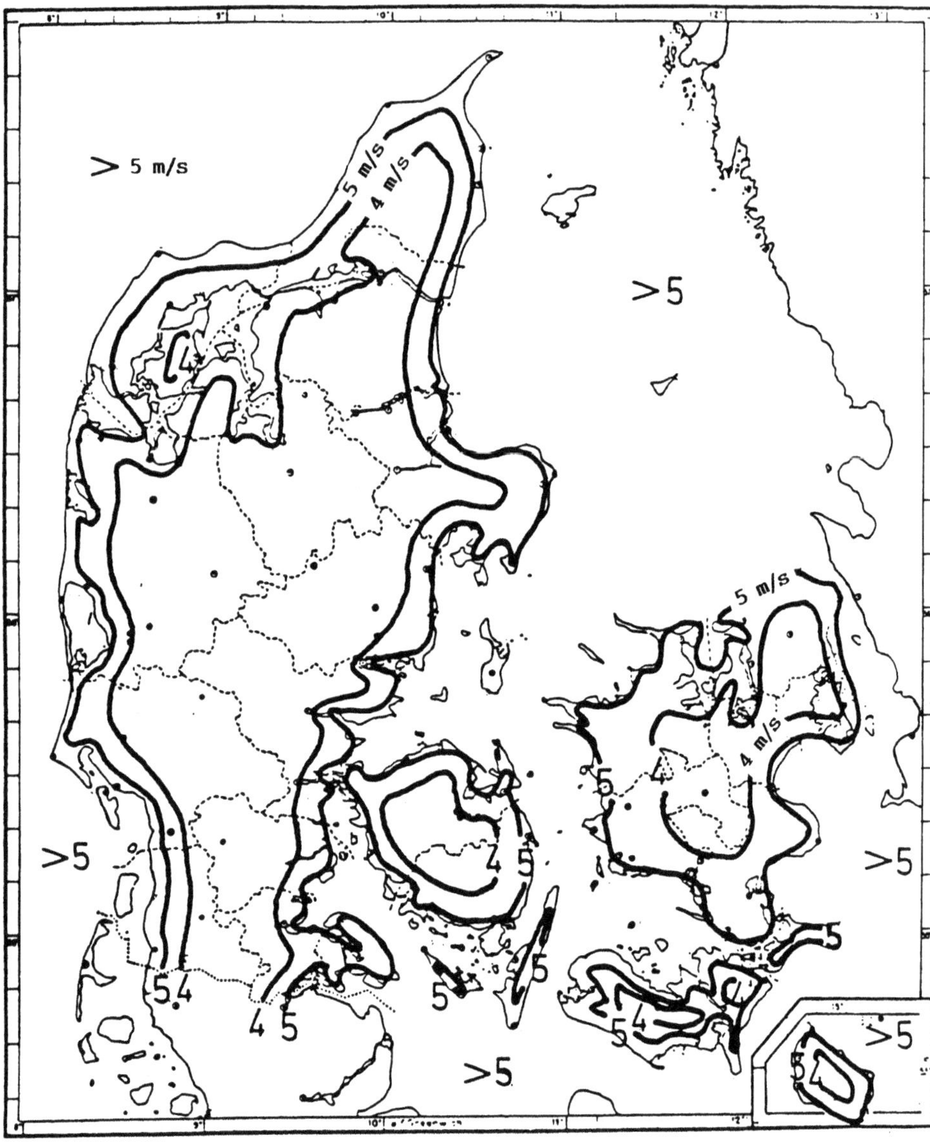

Fig. 2-5: 10 m Isovent Map of DENMARK

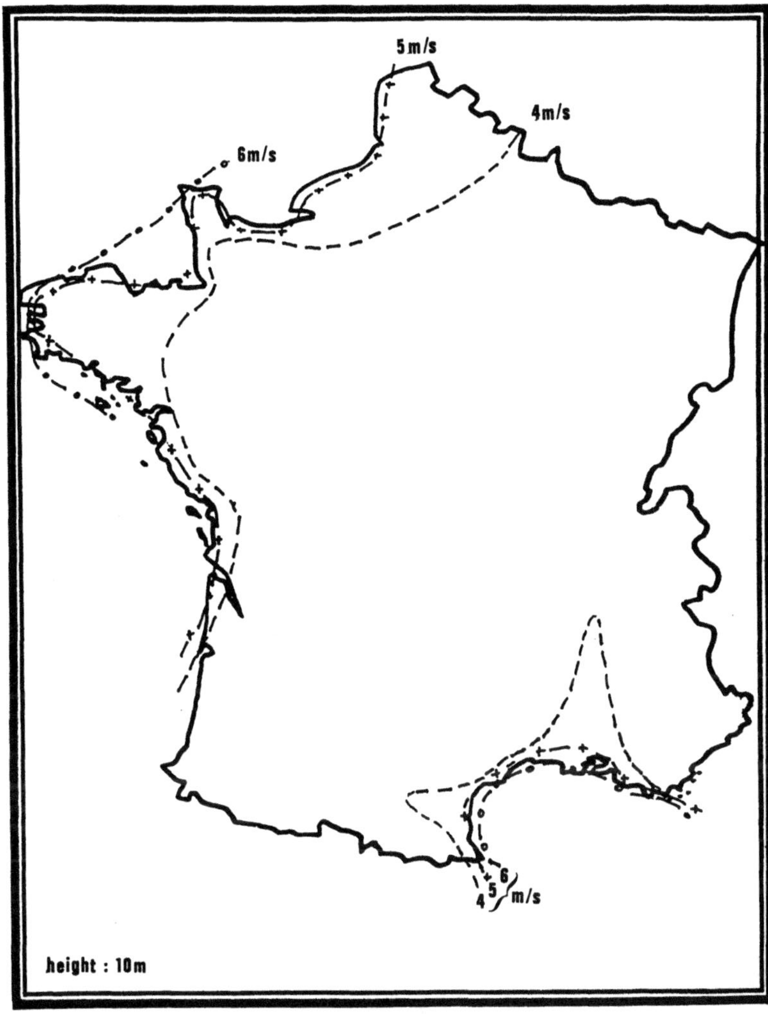

Source: Ph. DUCHENE - MARULLAZ, CSTB Nantes.

Fig. 2-6: 10 m Isovent Map of FRANCE (over 4 m/s)

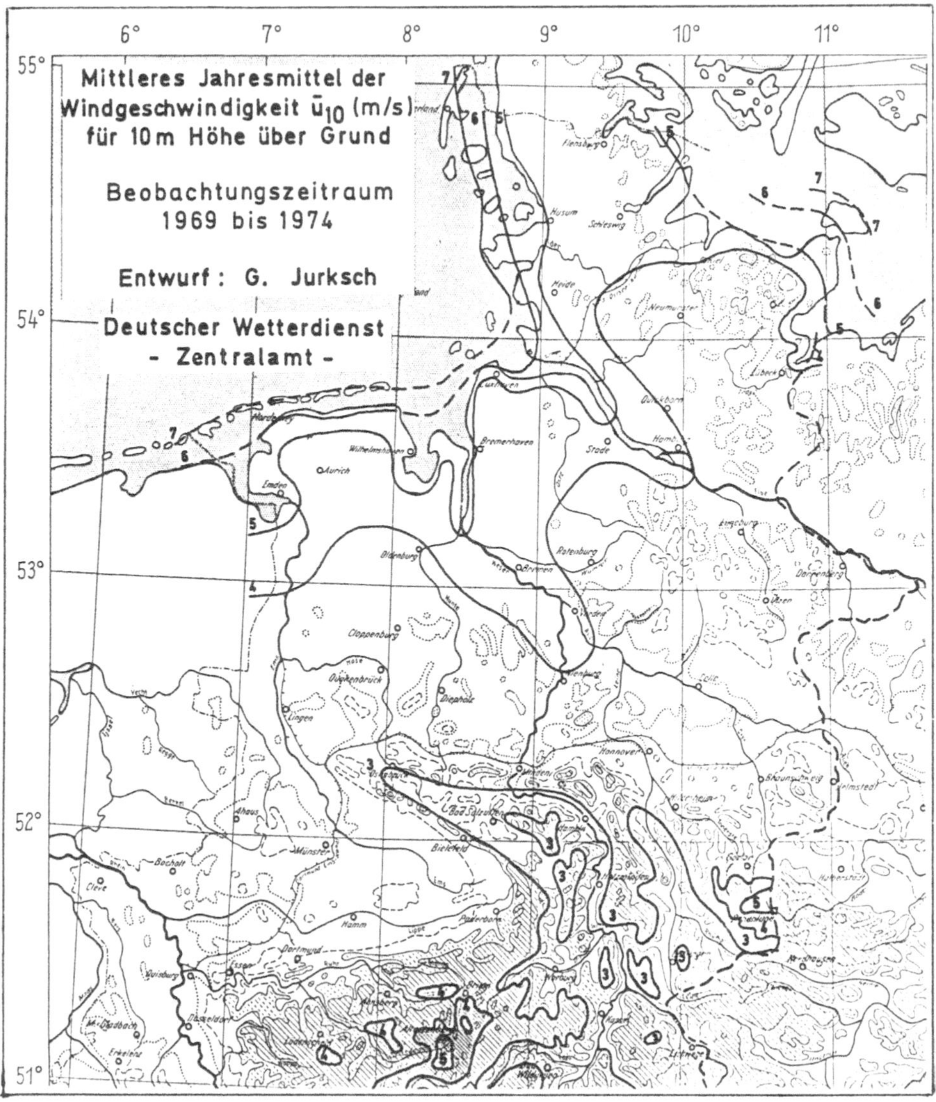

Fig. 2-7: 10 m Isovent Map of WEST GERMANY

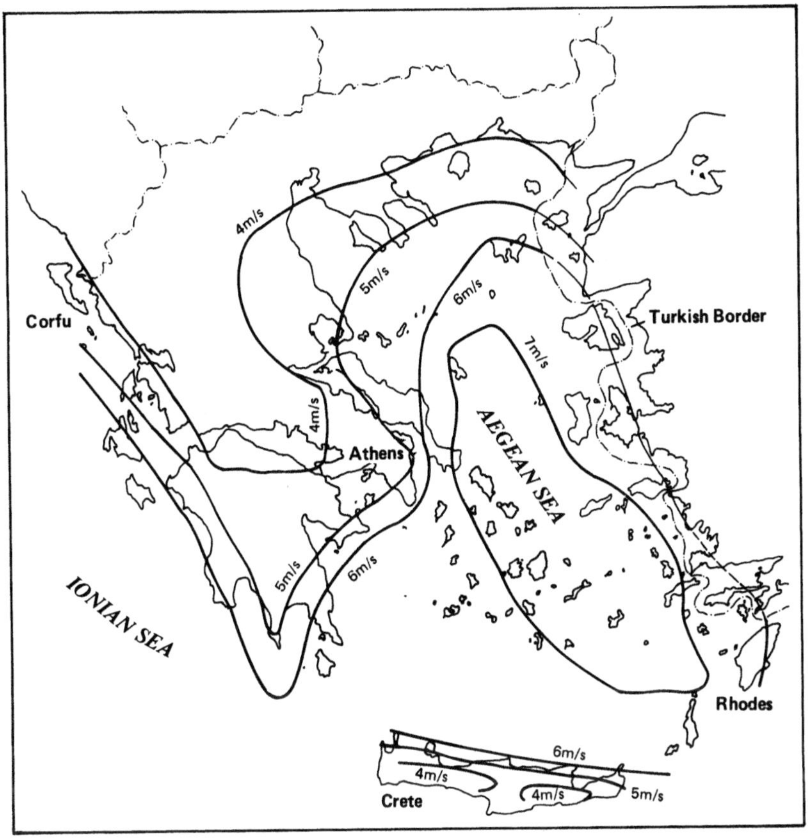

Fig. 2-8: 10 Isovent Map of GREECE

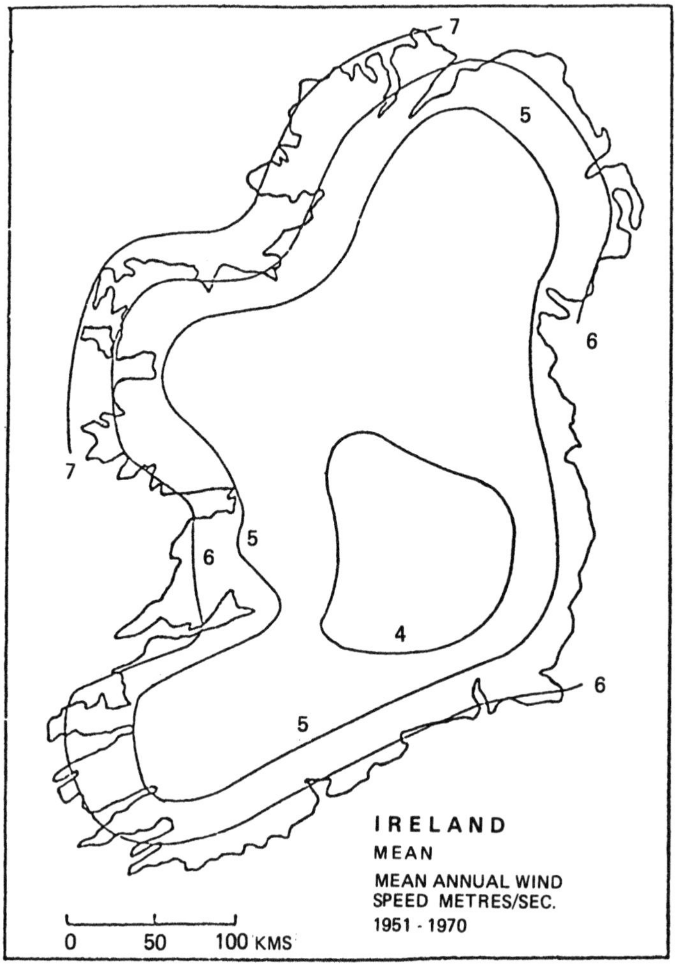

Fig. 2-9: 10 m Isovent Map of IRELAND

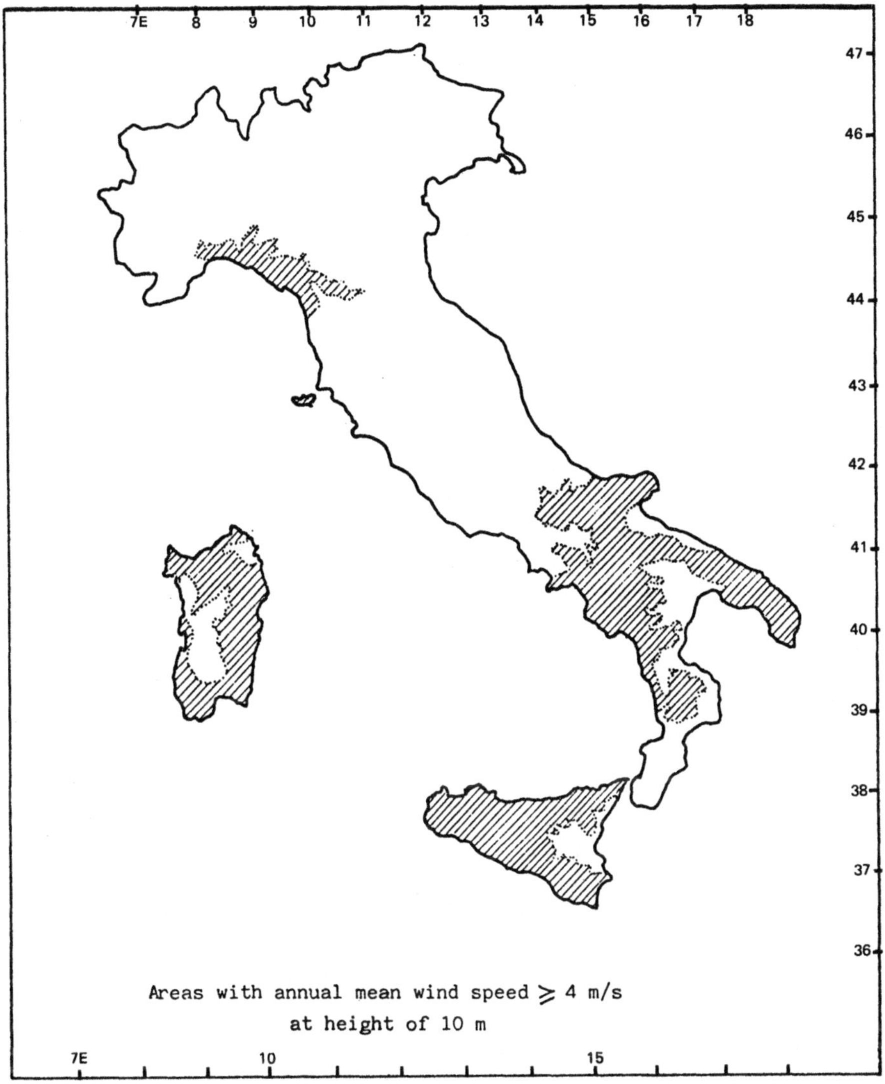

Fig. 2-10: Coastal Regions with Wind Speeds greater than 4 m/sec in ITALY

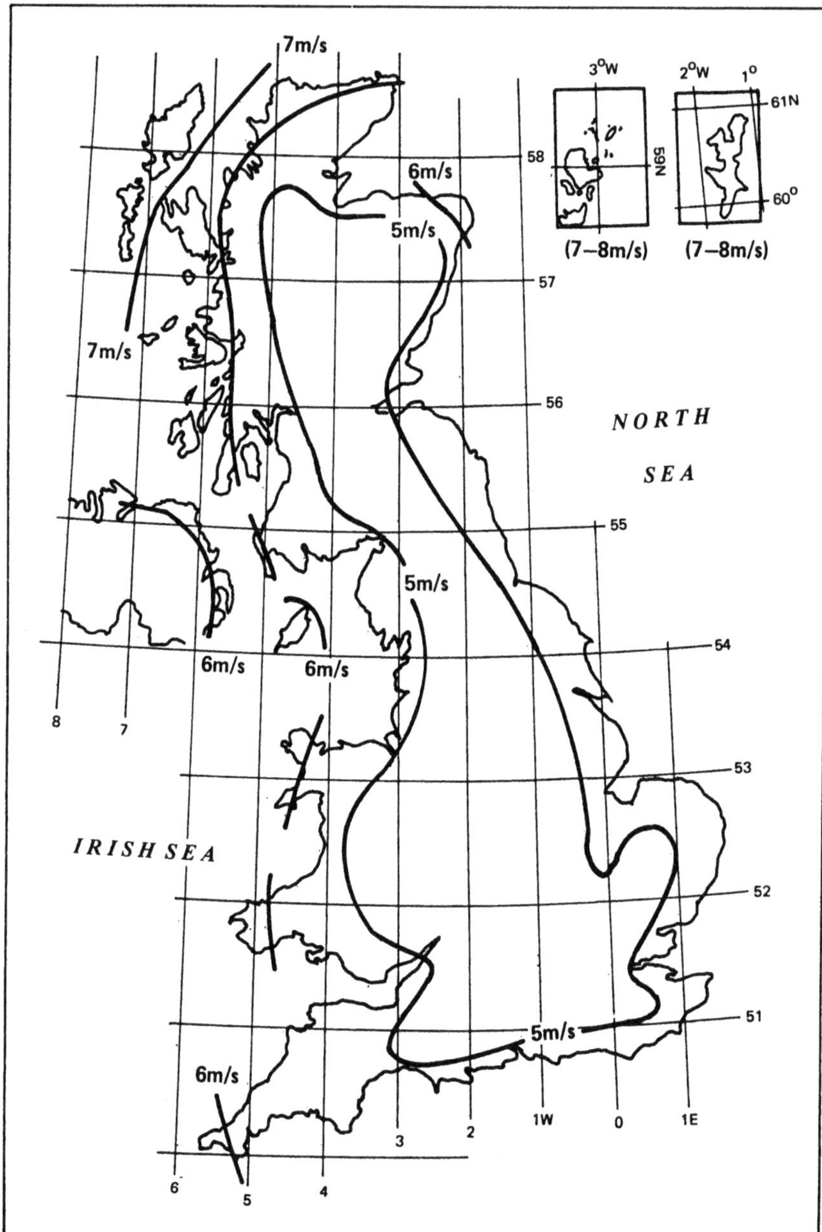

Fig. 2-11: 10 m Isovent Map of UNITED KINGDOM

2.2 Evaluation of the Sites and Conversion Potential

Having established the wind-rich areas, the actual sites suited to, and acceptable for the installation of wind turbines have to be decided upon. This task involves finding a method for surveying 1,658,000 km² within the scope of the time frame and workload of this study and producing meaningful results which can in turn be cross-checked and updated.

2.2.1 Description of the Method Used

The procedure developed was that of using a very detailed map, scale 1:100,000 and identifying all those areas which are excluded from consideration for the installation of large WECs. A net containing triangular grid points 5 km distant from one another was then superposed on the map. The grid points suited to WEC sites were counted and then processed. The details of the method are described in Appendix 2.

The following areas have not been investigated:

- Areas outside the national boundaries of the country considered.
- Areas in the offshore region with a depth more than -10 m below sea level
- Areas on land with ground levels higher than 1000 m altitude above sea level.
- Areas with yearly average wind velocities below 4 m/s (i.e. outside the area bounded by the isovent line with a mean velocity of 4 m/s).
- Areas outside the coastal region bounded by the isovent line with a mean velocity of 4 m/s first encountered when proceeding inland from the coast.

To achieve comparability, the application of defined and controllable exclusion rules proved necessary. Subjective measures like "unlikeliness" and "visual pollution" have not in theory influenced the results, although in practice they may well do so. However, the extent of this kind of restriction is subject to personal judgement and can be taken into account, by certain personal reduction factors.

The number of grid points for eligible sites of large wind energy converters in the subfield should be determined by considering the following restrictions:

- physical restrictions:
  - waterways
  - lake areas with depth of water greater than 10 m
  - traffic areas
  - airports
  - mountainous areas with slopes steeper than 10 %
  - forests
  - areas with no firm ground (moorland, quicksands)
- legal restrictions:
  - residential areas
  - industrial areas
  - nature reserves
  - air lanes (low level, near airports)
  - military prohibited areas (proving grounds, low level flight ranges)
  - areas used for coastal protection purposes (dikes, sand, dunes)
  - areas currently in use for surface mining
- political restrictions:
  - areas for which legal restrictions as mentioned above are foreseen and will be realized in the near future

The question that was open at the beginning of this study concerning the accuracy of the method applied has since been answered with the aid of results from the Dutch and Danish national programmes dealing with very detailed siting studies which have become available. The agreement was surprisingly small, in the order of 20 %, and the over-optimistic values can be correlated to the different exclusion rules applied.

From both the Danish and Dutch comparisons it might be concluded that the Action 1 approach leads to an overestimation of the land use factor, but that the final answer gives a reasonable estimate of the upper limit of the technical wind energy potential. This limit is so high that it will not be reached because of other technical, economic and political limitations not considered in the evaluation.

As to the potential sites for small WECs, the assumption was made that small WECs will generally be operated by private owners in Europe. Hence the number of houses not connected to the grid or in remote locations was of interest. This definition focuses on a possible market potential instead of the physical potential, because the great power variation within the category "small WEC" - ranging from 1 kW to about 100 kW - and the different energy requirements of potential customers do not permit better than order-of-magnitude figures to be obtained. On the other hand, most of the European electricity generating boards have already indicated that they are not interested in forms of small WECs as a solution to large-scale electricity production. Even so, the evaluation of houses suited to the operation of a small WEC remains rather inaccurate for the reason of access to the many variables. It must therefore be emphasized that an attempt was made to derive meaningful data in the order of magnitude but that these data are open to further refinements.

2.2.2 Results of the Analysis

The division into geographical subfields allows a good control and comparison of the data supplied by the national contractors. The evaluation is based on a statistical quantity of approximately 2000 grid points, 5 km distant from one another, per country on average (500 points in Belgium was the lowest, and 3 300 points in Italy the highest number). Despite these large quantities, accidental errors cannot be excluded in any statistical method, especially at boundaries. Hence, it should be noted that this siting analysis will give general averaged values and will not supersede special detailed planning investigations in all cases of WEC installation.

The input data and results have been compiled in output tables, and the most important results have been summarized for each country in the tables to be published as a special CEC-report.

Potential for Small WECs

The data for Denmark, France, Italy and the Netherlands have been taken from statistical sources, those for Belgium and Luxembourg, West Germany, Greece, Ireland and the United Kingdom have been derived from maps as described in App. 2. The reasoning for the data is included in the national sections of App. 3.

The total quantity of remote houses suited to the installation of small WECs may be in the order of 560,000, as shown in Table 2-2. The attempt to derive a fraction to be quoted as a realistic potential in the sense of a future market was unsuccessful. Even the consideration of a missing grid connection did not help. The best approach could be found by following the national goals of the Dutch and Danish programmes for the installation of 15,000 and 60,000 small WECs respectively until the year 2000. Along this line, the contractors estimated the potential of the other countries. Thus a total of 145,000 small WEC sites resulted, as presented in Table 2-2.

It should be stressed that this procedure is as inaccurate as any other approach to quantify the customer of an undefined market. The quantity of remote houses might serve as a guideline, but in the case of the United Kingdom an excessive quantity resulted owing to the fact that the whole country is exposed to wind speeds of more than 4 m/s - in general. Detailed mapping would certainly show many areas of reduced wind speeds in the 10 m layer and thus reduce the number of sites. But this procedure could not be followed in this study as it calls for a much higher degree of measurements and evaluation. In order to balance, the quantity of probable customers for the U. K. was assumed to be of the same order as for the other countries.

Table 2-2:
Summary of the Potential Sites for Small WECs within the Coastline of the EC-Countries

Source	Country	Total Potential	Not Connected to Grid	National Goals or Probable Customers	Land Area (km^2)
Statistical sources	Denmark	100 000	-	60 000	12 000
	France	50 000	10 000	10 000	83 049
	Italy	20 000	20 000	10 000	73 198
	Netherlands	100 000	-	15 000	13 730
Maps	Belgium/Lux.	25 000	-	8 000	11 388
	West Germany	35 000	-	10 000	32 436
	Greece	40 000	5 000	10 000	44 514
	Ireland	12 000	200	12 000	49 340
	United Kingdom*	278 200	1 000	10 000	239 969
	Total	560 000	36 200	145 000	559 826

145 000 x 30 MWh/a per unit
= 4.4 TWh/a

*For U. K., see remarks above

It must be pointed out that the result shown in Table 2-2 is wide open to discussion, as it should be the task of a market analysis to investigate the customers' attitudes. From a technical point of view, no rule was found to determine what fraction of the total potential would be realistic. Even the approach based on the distribution according to mean annual wind speeds can be used for reference purposes only since the actual wind speed in the lower class depends on local influences.

The assessment of the sites for small WECs is based on the assumption of private owners. Large-scale applications by clusters for centralized generation purposes has been excluded because the location of wind farms require a very detailed and extensive siting analysis and this would exceed the scope of the present study.

If one wishes to find the total potential of dispersed sites for small WECs, the statistical method developed contains the information to derive at least the order of magnitude. Although the specifications had large WECs in mind, the land use factor can be used for small WECs as well. The potential land area is 0.64 x 559,826 km² = 358,288 km². Assuming a rotor with 15 m diameter (50 kW size) and a spacing of 10 diameters in triangular arrangement, 51 WECs could be placed per km². Hence the total potential for rotors of 15 m will be 18,272,720 sites. The other calculation starts from the number of large WEC sites (409,750) and assumes a spacing of 150 m instead of 1 km, resulting in a factor of 44.4. Hence 44.4 x 409 750 = 18,211,111 sites may exist for rotors with 15 m diameter. Remembering that the statistical method gave optimistic values to be reduced by 40 % the appropriate number may be approximately 10 million sites for small (15 m) WECs.

Such an interpretation of the assessment study goes far beyond any realistic application and it was performed to indicate the possibilities if special analysis of some land areas is to be started. Normally, the dense package of 50 WECs per km² will exclude the usufruct of the ground for most purposes, hence the large-scale utilization of wind energy cannot be seen in connection with small WECs, but this does not necessarily exclude the temporary advantages of clusters in special cases.

The total potential of small WEC sites has been discussed in the preceding section, as well as the probable number to be expected. The computer programme calculated the equivalent specific energy of all WEC sites and they are included in the tables of App. 3. However, it is irrelevant to consider the total potential of 560,000 units and therefore an energy breakdown will not be presented here.

To give the order to magnitude which might be contributed by small WECs, the probable potential of 150,000 units was taken together with an assumed rotor size of 15 kW in a medium wind class of about 5 m/s. This will lead to an annual energy of approximately 30 MWh per rotor, totalling approximately 4.5 TWh. It should be noted that the purpose of small WECs will not be to contribute to the national energy supply but to improve the living conditions in remote dwellings.

The installation of wind farms in the U.S.A. raised the discussion of small WEC clusters in Europe. It is likely that small WECs will be installed in groups locally, especially as they offer advantages in hardware price for the time being, but from an energy assessment point of view the overall energy contribution will be small. The main reason for this is the large numbers needed, the diagrams show that 75 small WECs (50 kW) must be used instead of 1 large WEC (3 MW) to produce the same annual energy. In view of the dense installation of 75 units on 0.86 km², one might expect additional costs for buying the entire land area, as it might no longer be used for other purposes.

To conclude, it seems likely that small WECs will serve private users rather than public utility companies, except in remote areas and island networks with limited capacity. Despite this limitation, the potential number of users derived in this study is still high and will justify an increased involvement of industry.

On-Land Potential for Large WECs

Note: A hypothetical rotor size of 100 m diameter, with a spacing between WECs of 10 times the diameter was chosen. For divergent spacing the appropriate correction factor should be applied.

The results according to the work instructions and specified restrictions are summarized in Table 2-3, which shows the potential sites for large WECs and the annual energy for all countries classified by the wind classes at 10 m and 100 m height. The values for Denmark and the Netherlands have been taken from the national studies and assimilated to the statistical method. The values for the United Kingdom and France stem from a 10 km triangular grid instead of a 5 km grid.

Table 2-3:
Potential On-Land Sites for Large WECs and Potential Annual Energy

on-land	number of large WECs in a 1 km grid $\bar{v}_{100}$ classes in m/s						number of large WECs in 1 km grid $\bar{v}_{10}$ classes						area in km²					
	< 7	7-8	8-9	9-10	> 10	Σ	4-5*	5-6	6-7	7-8	8-9	Σ	4-5	5-6	6-7	7-8	8-9	Σ
Belgium/Lux.	7825	1500	---	---	---	9325	7825	1500	---	---	---	9325	9937	1451	---	---	---	11388
Denmark	8343	3094	198	---	---	11635	8343	3094	198	---		11635	7225	2680	172	---	---	10077
France	---	44500	5500	100	---	50100	44500	5500	100	---	---	50100	73610	9266	173	---	---	83049
W. Germany	---	8550	2025	450	25	11050	8550	2025	450	25	---	11050	16929	3594	736	152	---	21411
U. Kingdom	65400	68600	39000	2600	4100	179700	87000	78800	9800	4100	---	179700	116910	103227	14549	5283	---	239969
Greece	23900	---	12525	7025	3025	46475	23900	12525	7025	3025	---	46475	22841	12024	6945	2706	---	44516
Ireland	39425	7750	875	150	---	48220	18925	15400	10800	2950	150	48225	18965	17125	10457	2663	130	49340
Italy	37400	2050	2725	1075	75	43325	39450	2725	1150	---	---	43325	66097	5109	1992	---	---	73198
Netherlands	600	7980	1400	---	---	9980	600	7980	1400	---	---	9980	----	---	----	---	----	13730
total	182893	144024	64248	11400	7225	409810	239093	129549	30923	10075	150	409810	(325289)	(151796)	(34852)	(10804)	(130)	557703

	annual energy in TWh/a						annual energy in TWh/a						land use factor					
Belgium/Lux.	65	17	---	---	---	82	65	17	---	---	---	82	0.68	0.89	---	---	---	0.71
Denmark	69	32	2	---	---	103	69	32	2	---	---	103	---	---	---	---	---	0.82
France	---	489	73	2	---	563	489	73	2	---	---	563	0.52	0.51	0.50	---	---	0.52
W. Germany	---	94	26	7	0.44	128	94	26	7	0.44	---	127	0.44	0.49	0.53	0.14	---	0.45
U. Kingdom	505	677	471	38	70	1760	718	848	125	70	---	1760	0.64	0.66	0.58	0.67	---	0.65
Greece	220	---	154	112	62	549	220	154	112	62	---	549	0.91	0.90	0.88	0.97	---	0.91
Ireland	298	84	12	2	---	395	125	123	109	36	2	395	0.86	0.78	0.89	0.96	1.00	0.84
Italy	288	22	35	17	1	363	309	35	19	---	---	363	0.52	0.46	0.50	---	---	0.52
Netherland	7	90	21	---	---	118	7	90	21	---	---	118	---	---	---	---	---	0.63
total	1451	1504	794	178	134	4061	2096*	1398	396	169	2	4061	---	---	---	---	---	0.64

* large scale wind energy utilization will be questionable in the range $\bar{V}_{10}$ = 4-5 m/s.

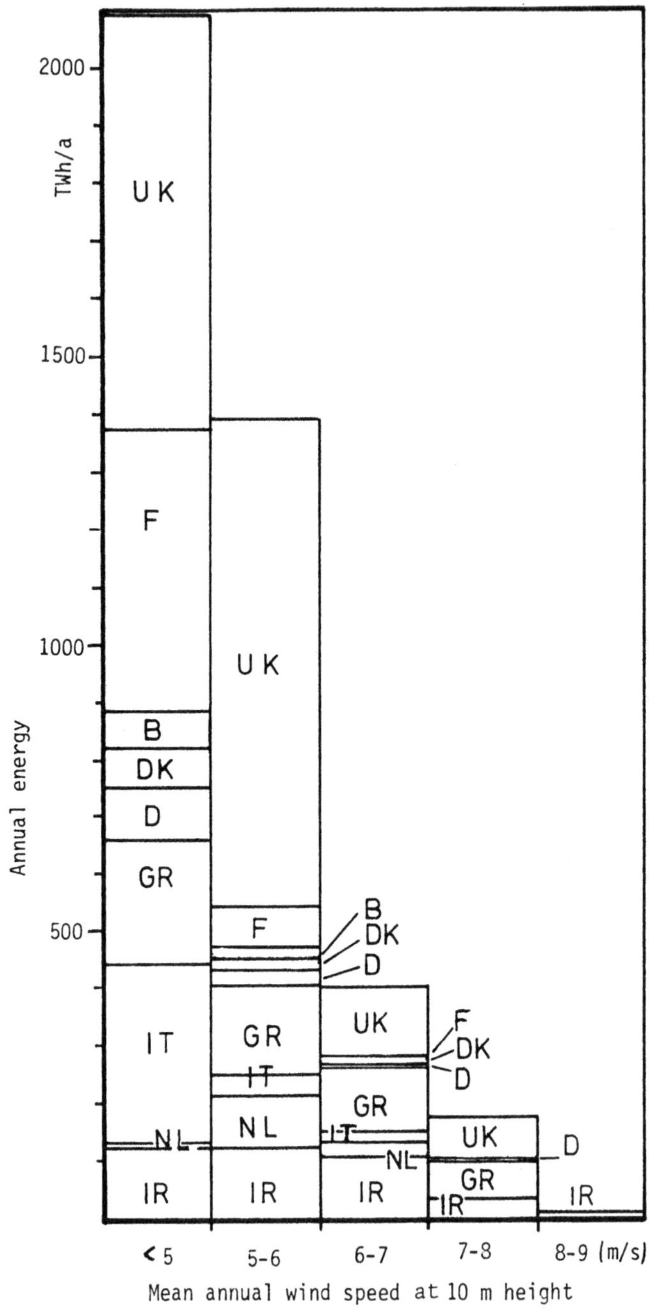

Figure 2-12:
The Potential Energy Contribution from the Various Wind Zones

The total quantity of potential sites ranges in the order of 400 000, dispersed on approximately 350,000 km² of land out of approximately 550 000 km², giving a land use factor of 64 %. The 400,000 sites have a potential energy production of more than 4000 TWh/a which vastly exceeds the consumption of 1214 TWh in 1980 in EUROPE-10.

Nationally, these technical potentials exceed the national electricity consumption several times, except in West Germany, so that areas of low wind speeds can be neglected.

The distribution according to the various wind speeds at 100 m and 10 m height respectively shows the differences in the wind speed calculations mentioned in section 2.1. When the figures in the high V_{10} and V_{100} class are compared, inconsistent roughness factors between the countries can be noticed, supporting the necessity of more investigations in wind mapping and processing as is already being done by Action 2.

The study shows that all countries except Belgium possess several 100 to 1000 sites in the high wind speed regime, an encouraging finding for the start of an energy utilization programme. The details of the siting results will be discussed in the national section of the ECE-report and in Section 2.4 below dealing with the energy potential.

The wind speed range 4-5 m/s was included in this study to cover all areas of potential interest. Except for some special sites, it is questionable whether such areas could be used for a large-scale wind energy utilization.

Off-shore Potential for Large WECs

This part of the study gave rise to intense discussions among the various research groups owing to the different definitions of the off-shore area. Very careful studies performed in the Netherlands [2-6], the United Kingdom [2-7] and Denmark have selected different boundary conditions. They will be discussed briefly in the national sections of this chapter.

The principal concern of the present study was comprehension of the various influences. Land-based wind turbines are well understood as far as technical and economic matters are concerned. Shifting the foundation line from on-shore to shallow waters of several metres depth will cause extra problems and expenses but these are accessible to calculations. Going to the water depth of half the tower height or more would mean a complete new engineering of the total system due to the new environment. But there are also strong reservations against constructions in the 0-5 m depth due to the danger of underwash.

In view of the many arguments, wind turbine sites in shallow waters to 10 m depth have been evaluated within the scope of this study. The results, classified according to wind speeds at 10 m and 100 m height, are presented in the upper part of Table 2-4.

Greece and Ireland have rocky coastlines, hence an assessment will produce in meaningless figures. In the Netherlands the existing results of a study for a closed off-shore windpark - outside 10 m water depth - led to the conclusion that no suitable sites exist in shallow waters.

A total number of approximately 28,000 potential sites could be found within the 10 m waters of the European coastline, equivalent to an annual energy production capacity of approximately 360 TWh. It may be concluded that the real potential is far greater if water depths of more than 10 m are included.

Whereas it was difficult to pay attention to the restrictions on land, the situation in shallow waters is even more complex as it is likely that limitations will come up as soon as large-scale utilization of these areas is foreseen.

Table 2-4:
Potential Off-Shore Sites for Large WECs within Waters of 10 m Depth and Potential Annual Energy

off-shore	number of large WECs in a 1 km grid $\bar{v}_{100}$ classes in m/s						number of large WECs in 1 km grid $\bar{v}_{10}$ classes						area in km²					
	6-7	7-8	8-9	9-10	> 10	Σ	4-5*	5-6	6-7	7-8	8-9	Σ	4-5	5-6	6-7	7-8	8-9	Σ
Belgium/Lux.	---	---	---	575	---	575	---	---	575	---	---	575	---	---	736	---	---	736
Denmark	---	---	6190	---	---	6190	---	---	---	6190	---	6190	---	---	---	5360	---	5360
France	---	1200	4900	800	---	6900	1200	4900	800	---	---	6900	1039	4850	693	---	---	6582
W. Germany	---	175	1950	2450	2275	6850	175	1950	2450	2275	---	6850	152	1862	2382	2468	---	6864
U. Kingdom	---	---	5000	---	---	5000	---	5000	---	---	---	5000	---	5023	---	---	---	5023
Greece	---	---	---	---	---	---	---	---	---	---	---	---	---	---	---	---	---	---
Ireland	---	---	---	---	---	---	---	---	---	---	---	---	---	---	---	---	---	---
Italy	2250	---	---	---	---	2250	2250	---	---	---	---	2250	2013	---	---	---	---	2013
Netherlands	---	---	---	---	---	---	---	---	---	---	---	---	---	---	---	---	---	---
total	175	1375	18040	3825	2275	27765	3625	11850	3825	8465	---	27765	3204	11735	3811	14592	---	26578

	annual energy in TWh/a						annual energy in TWh/a						area use factor					
Belgium/Lux.	---	---	---	8	---	8	---	---	---	---	---	---	---	---	0.68	---	---	0.68
Denmark	---	---	79	---	---	79	---	---	---	79	---	79	---	---	---	0.98	---	0.98
France	---	13	65	13	---	91	13	65	13	---	---	91	1.00	0.88	1.00	---	---	0.91
W. Germany	---	2	25	37	40	105	2	25	37	40	---	105	1.00	0.91	0.89	0.80	---	0.87
U. Kingdom	---	---	60	---	---	60	---	60	---	---	---	60	---	0.86	---	---	---	0.86
Greece	---	---	---	---	---	---	---	---	---	---	---	---	---	---	---	---	---	---
Ireland	---	---	---	---	---	---	---	---	---	---	---	---	---	---	---	---	---	---
Italy	2	---	---	---	---	15	15	---	---	---	---	15	0.97	---	---	---	---	0.97
Netherland	---	---	---	---	---	---	---	---	---	---	---	---	---	---	---	---	---	---
total	2	15	230	59	40	359	30*	151	59	119	---	359						0.92

* large-scale wind energy utilization is questionable in the range $\bar{V}_{10}$ = 4-5 m/s

The energy potential of the off-shore sites within 10 m water depth is shown in the lower part of Table 2-4. An energy equivalent of 360 TWh was deduced, which is rather low in comparison with the land-based sites. But it should be noted that this analysis covers only a small portion of the off-shore areas, and these figures serve to indicate the large total potential.

The installation costs off-shore will be noticeably higher than on land, however the wind class is better. In some cases, like West Germany, the large areas of sandbank covered by water only at high tide represent an excellent potential of about 26 TWh per year, if siting is accepted on these 1500 km^2.

2.3 Geographical Distribution of the Potential

For a better interpretation of what lies behind the tabulated values given above, accumulated figures have been mapped for the areas of interest in the countries analysed (Figs. 2-13 to 2-22). They represent the number of large WEC sites with 1 km spacing and the corresponding annual energy production in TWh/a. The equivalent power is found by assuming an installed capacity of 3 MW per wind turbine.

Although it is known that the mapped quantities are unrealistic, they represent the only safe ground as the result of the analysis and the application of reduction considerations must start from these quantities. For example, it is quite easy to exclude all areas with wind speeds below 5 m/s and to follow the impact for the country.

As mentioned before, Denmark and the Netherlands have performed more detailed siting analyses within their national programmes. Therefore, these results are presented here also.

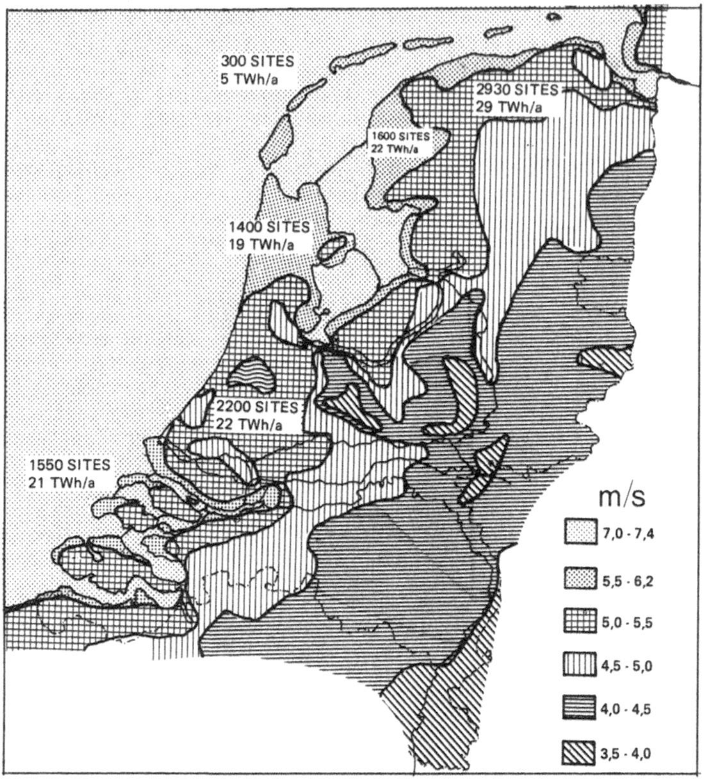

Fig. 2-13: Geographical Distribution of Potential Sites for Large WECs and Equivalent Annual Energy for the NETHERLANDS

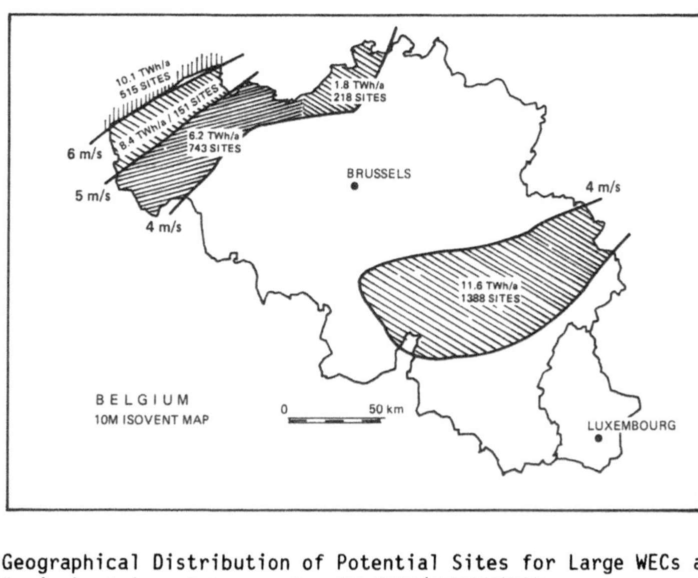

Fig. 2-14: Geographical Distribution of Potential Sites for Large WECs and Equivalent Annual Energy for BELGIUM/LUXEMBOURG

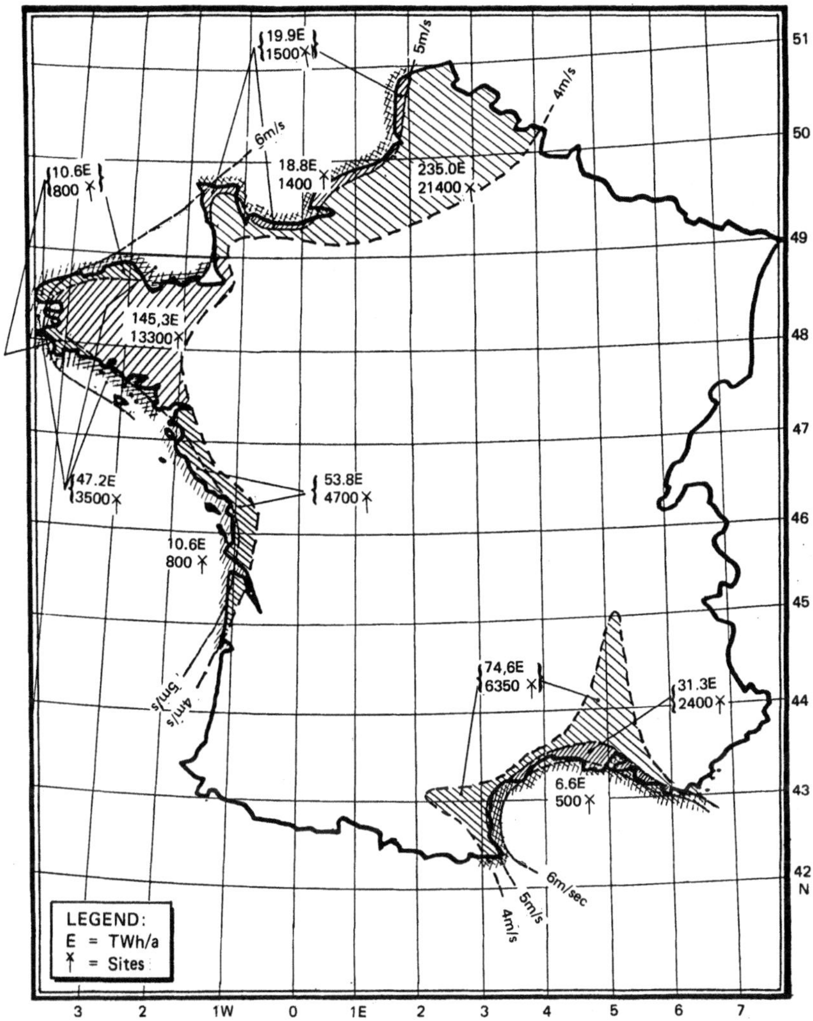

Fig. 2-15: Geographical Distribution of Potential Sites for large WECs and Equivalent Annual Energy Production for FRANCE

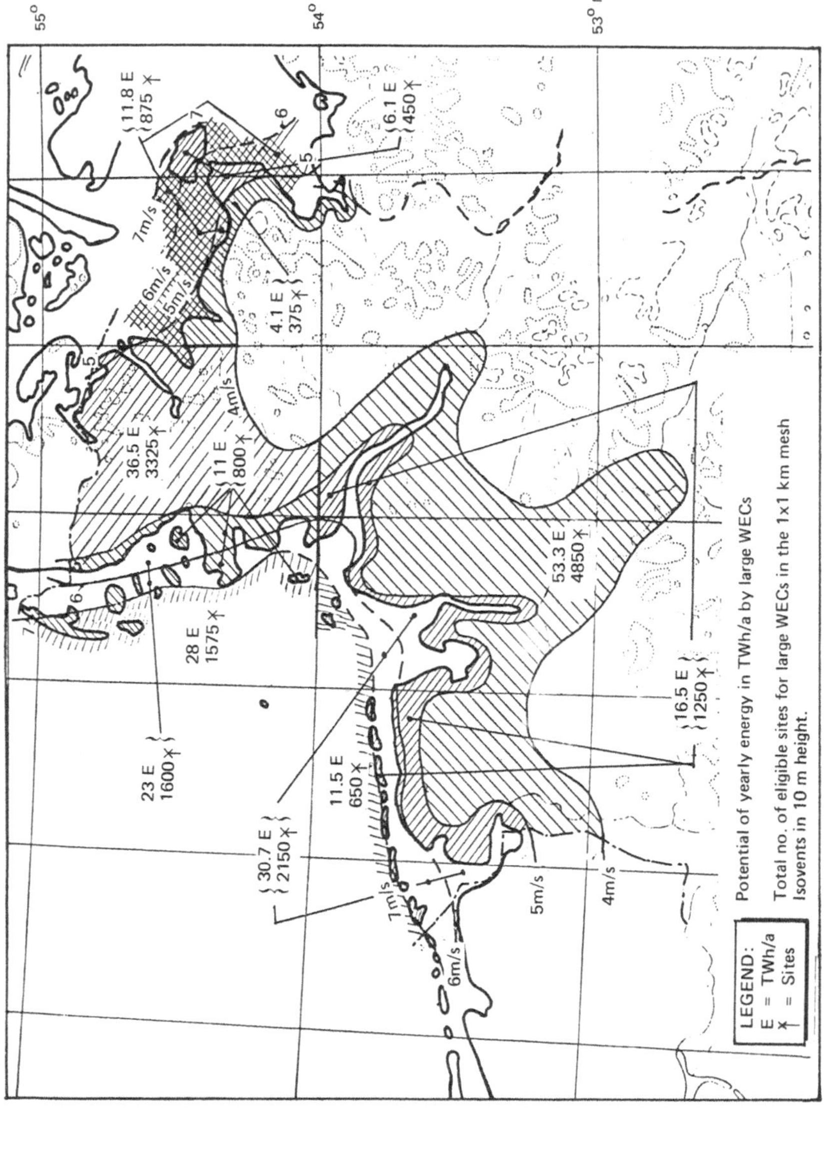

Fig. 2-16: Geographical Distribution of Potential Sites for large WECs and Equivalent Annual Energy Production for WEST GERMANY

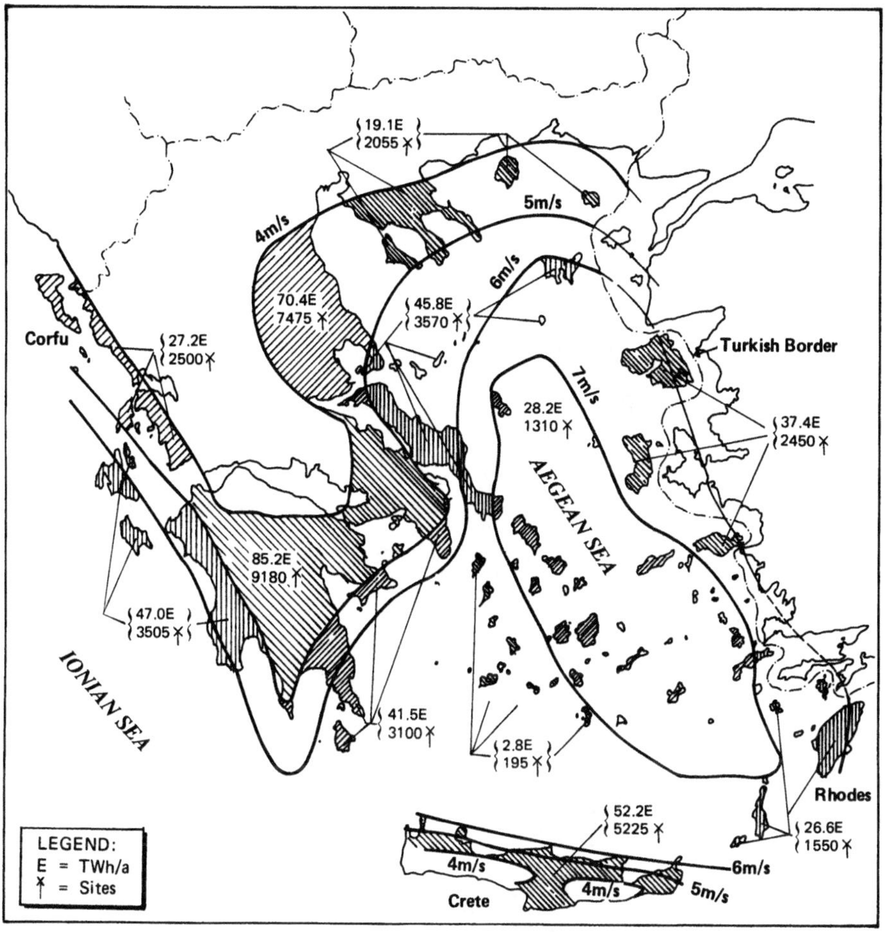

Fig. 2-17: Geographical Distribution of Potential Sites for Large WECs and Equivalent Annual Energy for GREECE

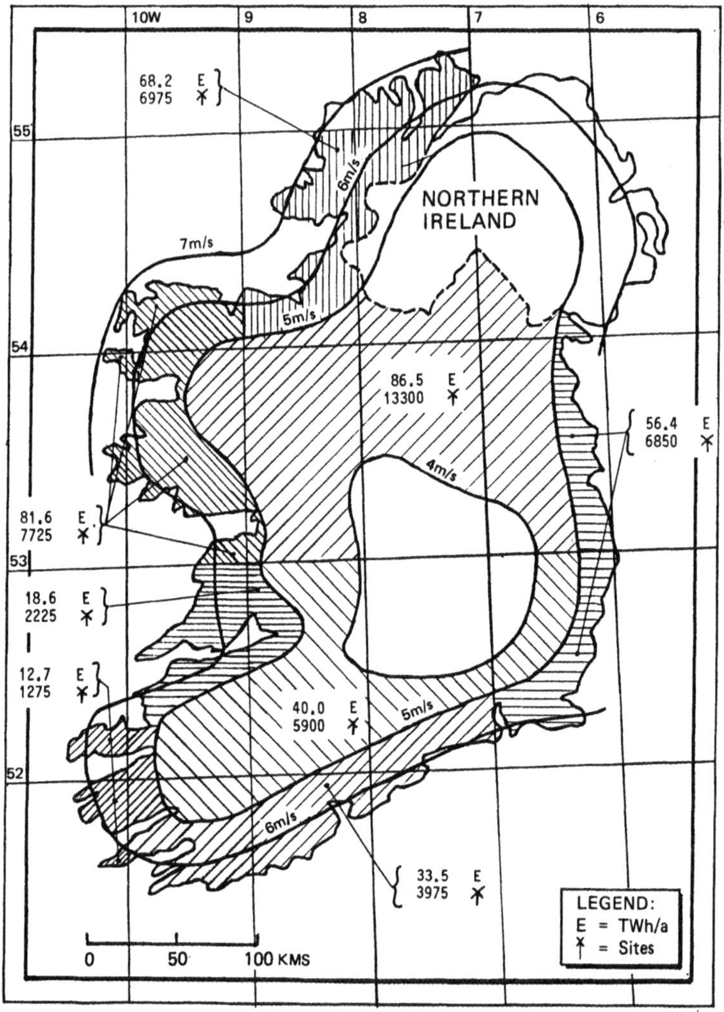

Fig. 2-18: Geographical Distribution of Potential Sites for large WECs and Equivalent Annual Energy Production for IRELAND

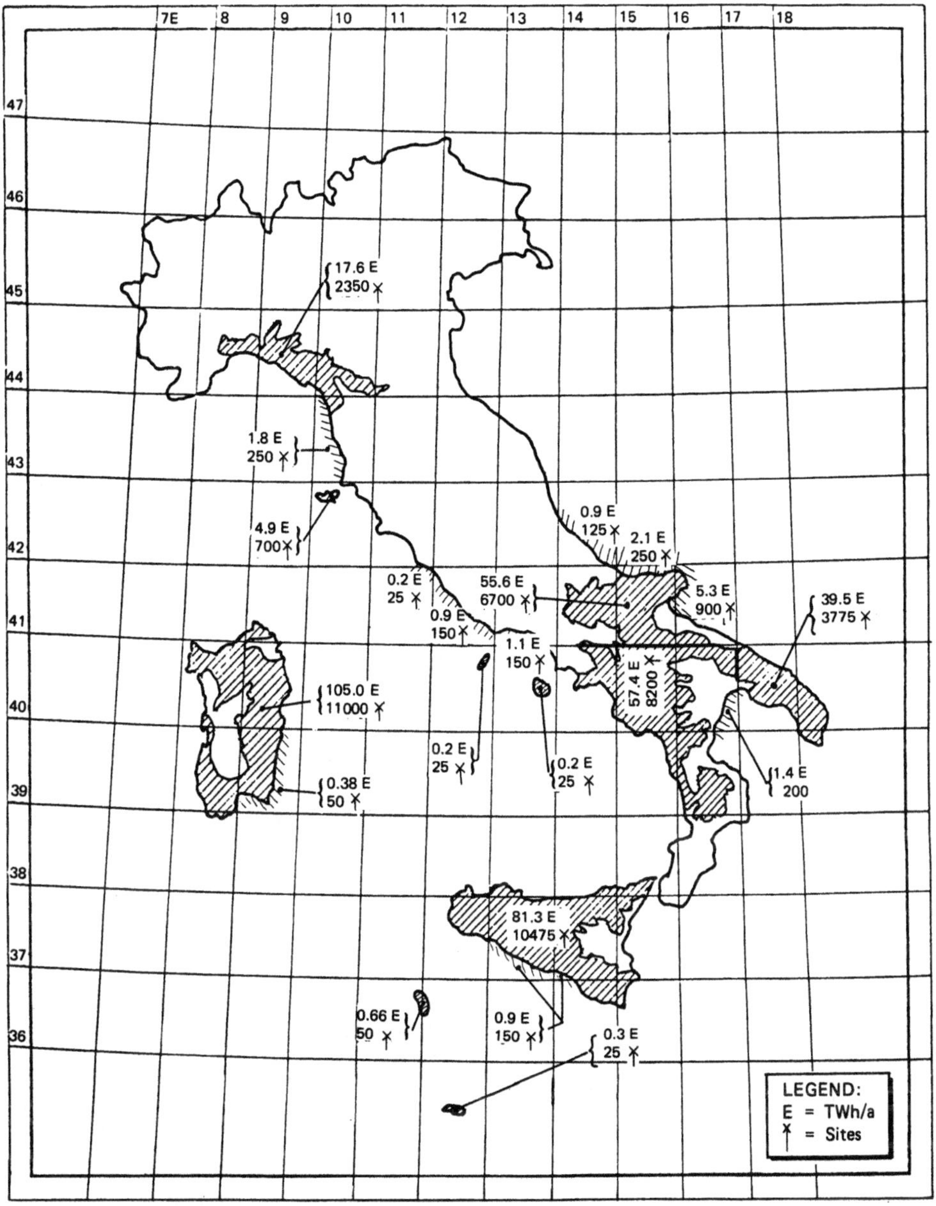

Fig. 2-19: Geographical Distribution of Potential Sites for large WECs and Equivalent Annual Energy Production for ITALY

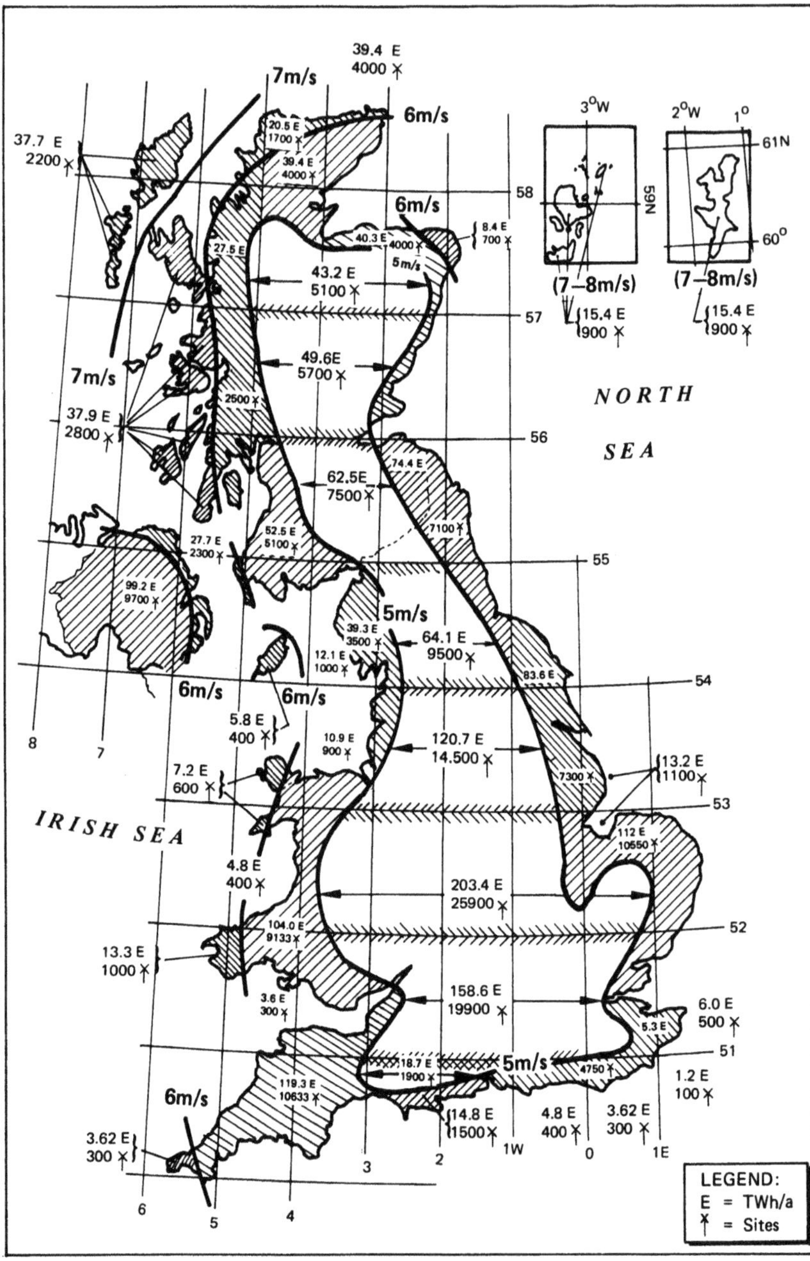

Fig. 2-20: Geographical Distribution of Potential Sites for large WECs and Equivalent Annual Energy Production for the UNITED KINGDOM

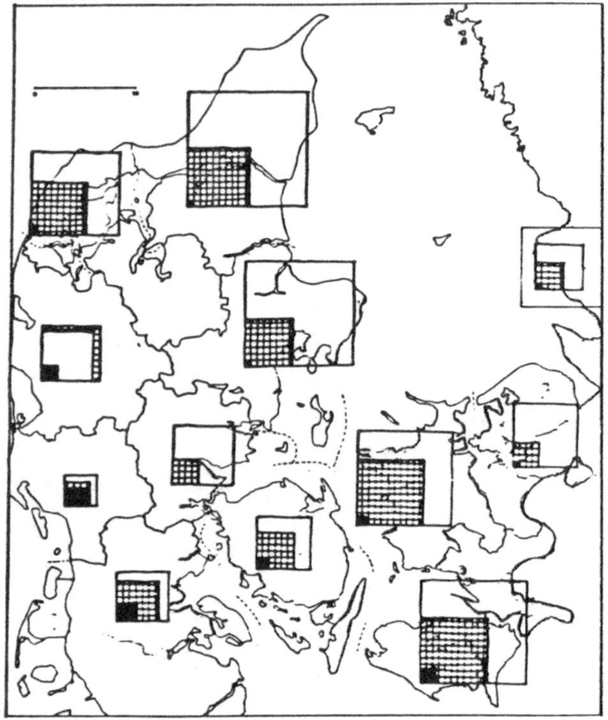

Fig. 2-21:
The distribution of the areas in the 3 best energy classes in model 2 for DENMARK

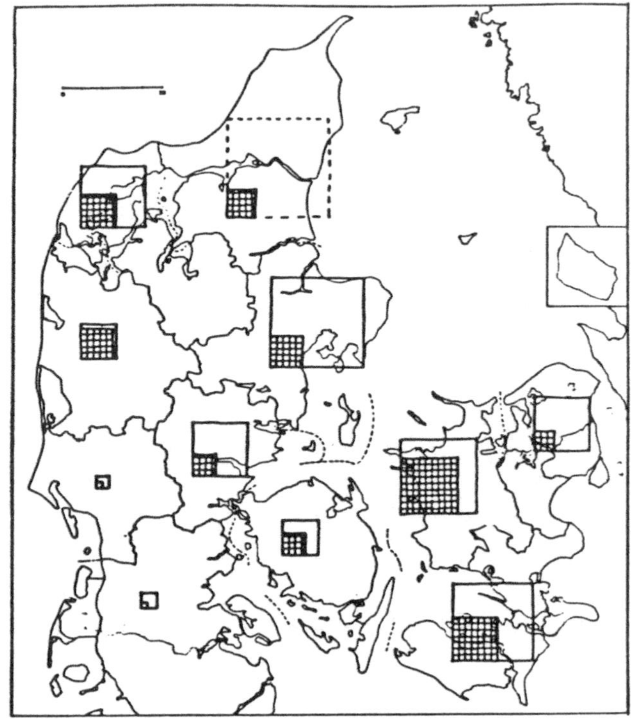

Fig. 2-22:
The distribution of the areas in the 3 best energy classes in model 5 for DENMARK

model 2		model 5
172 km^2	class A area	0 km^2
2680 km^2	class B area	1075 km^2
7225 km^2	class C area	3315 km^2
10077 km^2		4390 km^2

2.4 Discussion of the Siting and Conversion Potential

A good analysis of the influence of various restrictions was presented by Danish researchers; thus the available area of 12 230 km² in Model 1 diminished to 4390 km² in Model 5, 100 % to 36 % respectively (see App. 2). The Dutch investigation concluded that the specifications for siting in the Action 1 study would lead to a land-use factor of 63 % instead of 38 % resulting from the Dutch assumptions, meaning a reduction by 40 %. The same result was reported by an Italian cross-check and comparison of both paths.

In view of the many individual differences between countries, a land-use factor of 30 % to 40 % seems acceptable for most cases, with a higher factor for the less populated countries. The main reason for these differences in the land-use factor is the individual estimate of several "unlikelihoods" in the site selection.

Some systematic tendencies for overestimation were noticed during the evaluation. First of all, the quality and age of the maps should be mentioned. The better the information on the map, the more accurate was the analysis.

The spacing of the grid applied influenced the decision in so far as a short spacing of 1 km in the analysis led to more rejections than a wide spacing. Sloping areas raise important questions, because the total surroundings should be taken into account with respect to shadow effects, prevailing directions, etc. The determination of a minimum distance from buildings and traffic lines will also influence the results.

At present, criteria for safety distances are still under discussion since results from the prototypes tested have not yet been evaluated in this respect. As soon as series installations are foreseen - as in the case of small WECs - safety criterial will be essential for the building permit.

Another factor not taken into account may result from a possible request of the Power Boards to utilize wind farms only, and not dispersed installations.

There are many influences beyond the rules adopted for this analysis which may reduce the land-use factor. But all the effects noticed still leave more than 30 % to be used for installations - so long as a strong public or political will against wind energy does not emerge.

It has been mentioned before that the specifications for the site selection enabled us to assess the potential with an accuracy of about 20 to 40 % too high. In particular the localization of the isovent lines was questionable and served mainly as a border for the subfields. The wind data for the 100 m level showed national differences to be corrected in the Wind Atlas of Action 2. For the time being this uncertainty could mean a shift of the site quantities from one wind speed class to the next, with the implication of altering the energy output by up to 20 %. With respect to the wind energy potential in most countries, these considerations can be regarded as minor effects in comparison with what might be accepted by the public or could be utilized by the electricity networks.

To gain a better understanding of the implications of these figures, Table 2-5 compares the quantity of large WEC sites resulting from the site analysis and those sites required to supply 10 % of the national electricity consumption. With the exception of West Germany, the wind energy potential in most countries- according to the specifications of this study - is several times greater than the national electricity needs, even if very restrictive selection rules were applied. The supply of 10 % of the annual electricity of EUROPE-10 could be furnished by either:

- 8455 large WECs of 100 m diameter, if the sites in the highest wind zones on land are filled up first, or
- 9480 large WECs of 100 m diameter, if the sites are dispersed evenly over all on-land wind zones with more than 5 m/s at 10 m height.

	Large WEC Sites	Annual Energy (TWh/a)	Electricity consumed in 1980 (TWh/a)	10 % of the annual consumption (TWh/a)	Quantity of WECs required if selection starts from highest wind class	Quantity of WECs required if selection is distributed over 5-9 m/s wind classes	Mean specific land area km^2/WEC
Belgium & Lux.	9325	82	51.2	5.1	450	450	3.2
Denmark	11635	103	24.3	2.4	150	230	12.4
France	50100	563	248.7	24.9	1850	1850	5.1
W. Germany	11050	128	351.4	35.1	2700	2500*	1.8
Greece	46475	549	21.7	2.2	110	160	135.5
Ireland	48220	398	9.7	1.0	75	110	276.1
Italy	43325	364	180.2	18.0	1150	1300	5.5
Netherlands	9980	118	62.5	6.3	420	530	26.9
Un. Kingdom	179700	1760	264.9	26.5	1550	2350	52.4
	409810	4061	1214.6	121.5	8455	9480	
	According to the specifications of the assessement study			Reduction to 10% of the national electricity needs, assuming 100 m diameter WEC			

* Valid for 33.4 TWh/a only not for 35.1 TWh/a

Table 2-5: Influence of Various Reduction Strategies on Land on Quantity of Large WEC Sites and Annual Energy Production

There is a difference in quantity of 10 % between the two assumptions and consequently cost differences in the same order. With respect to the generality of the assessment study, this difference is negligible when compared with the implications of more precise wind data.

The figures indicate the bandwidth of influence of site selection and do not include energy losses if clusters with reduced overall efficiencies are used. An interesting point arises from the mean specific land area per WEC if the units are evenly distributed over all zones with wind speeds higher than 5 m/s. These values indicate that parameters other than siting will probably become the limiting factor for wind utilization.

All the data for large WECs assume a hypothetical rotor of 100 m diameter. Deviations from this size would influence the number of rotors needed for the same energy production. The correct transformation can be done with the aid of the diagrams of Chapter 4 (Figs. 4-1 to 4-5) but a good approximation may be obtained by considering the swept area of the rotors to be compared. For instance, the supply of 10 % of the annual electricity of EUROPE-10 by 75 m diameter rotors would require 15,200 units if selection starts from the highest wind class, or 17,000 units if selection is distributed over 5-9 m/s wind classes - as composed with 8,455 and 9,480 units respectively for 100 m diameter rotors shown in Table 2-5. Thus the data must be modified according to the specific values of interest.

As regards the off-shore potential, the results of the high wind speed zones can be applied, showing the great attraction for future discussion. A detailed presentation of the various impacts would depend on too many assumptions and therefore would not improve the accuracy of this assessment study. Fundamentally, the situation is similar to the land-based WECs, and the forthcoming investigations and development work will make it possible to decide on the extent to which wind energy can be used in practice, both on-shore and off-shore.

3. ASSESSMENT OF THE UTILIZATION POTENTIAL

3.1 General Aspects

The preceding chapter outlined the high potential of the natural energy resource "wind" within the countries of Europe-10. It was also mentioned that unpredictable limitations could arise from land use conflicts which have affected even mature conversion technologies. Whereas this limitation can only be foreseen, without even a general quantification, the standard of the existing energy supply systems opens up another aspect for the evaluation of the probable usable potential. On one hand, the use of wind power might be limited by the transmission capacity of the electricity network from the areas with high winds; on the other hand, mixed operation of the power plants might cause intolerable losses in supply steadiness, quality etc. No limitation could be foreseen due to purely technical reasons, but limiting conditions have been found in the associated economics because wind energy plants will need back-up capacity from conventional plants, and acceptability is rated primarily by profitability in our economy.

Therefore, the results of the present investigation will have only a preliminary character for no operational experience of this type of plant mix is available, and it is to be expected that by the time large-scale wind energy utilization is introduced the costs for energy and advanced power stations will have increased. Nevertheless it is worth while to consider where the borderline might be placed.

3.2 Survey of the Existing Supply System and Electricity Consumption

The generation capacity of the European Community and its future prospects have been received by UNIPEDE [3-1]. The generation capacity existing in 1980 is shown in Table 3-1. A brief description of the main characteristics of the national systems is given in the supplementary CEC-report.

The peak demand and annual electricity consumption are summarized in Table 3-2. This energy is transmitted by a well developed network whose interconnection between the European countries is shown in Fig. 3-1. For the time being, this interlink is not used for the provision of reserve capacity, but the technical potential of such an existing installation should be borne in mind. The same is true for the capacities of the pumping, hydro and gas turbine power stations if more complex operational strategies have to be applied owing to the possible integration of wind energy.

Country	Nuclear	Thermal – monovalent: Coal	Thermal – monovalent: Brown coal	Thermal – monovalent: Oil	Thermal – monovalent: Natural gas or by-prod.	Thermal – polyvalent: with coal	Thermal – polyvalent: without coal	Hydro	Pumping and mixed	Gas-turbine, Diesel etc.	Not specified	Total
B	1666	155	-	1133	164	2282	3634	75	1056	839	-	11004
D	8607	14642	12997	12035	11611	12251	-	2666	3785	3450	726	82770
DK	-	440	-	1793	-	4170	-	-	-	265	-	6668
F	14394	7329	200	13060	800	6000	1100	17671	1614	543	596	63307
GR	-	-	2083	1429	-	-	-	1414	-	376	30	5332
I	1424	-	61	15423	470	4626	7097	12170	3650	1465	434	46820
Irl	-	14	413	1608	369	-	85	220	292	-	-	3001
NL	498	-	-	617	1650	1641	10582	-	-	373	162	15523
UK	6458	40141	-	12240	-	4266	-	1391	1060	3553	4534	73643
EUR-10	33047	62721	15754	59338	15064	35236	22498	35607	11457	10864	6482	308068
	11 %	20 %	5 %	19 %	5 %	11 %	7 %	12 %	4 %	4 %	2 %	100 %

Notes

*Abbreviations for names of countries are explained in the Glossary (see "Europe-10").

Netherlands: the breakdown concerns public services only.

United Kingdom: the breakdown between fuel categories for industrial producers (about 4500 MW) is unreliable, consequently the corresponding capacities have been listed under the "Not specified" category.

Table 3-1: Position of Plant Mix in Europe-10 on 31 December 1980 (net MW)

	PEAK DEMAND [GW]			ANNUAL CONSUMPTION OF ELECTRICITY [TWh]		
	1979	1980	1981	1979	1980	1981
EUR-10	216.2	220.4	222.0 219.0	1207.1	1214.6	1265.4 1249.6
B	7.8	7.9	7.9	48.0	47.7	47.7
D	60.6	61.4	66.0	348.9	351.4	359.0
DK	4.7	4.7	5.1 5.0	24.2	24.3	24.6 24.1
F	40.9	44.1	45.0	235.6	248.7	260.0 256.0
GR	3.3	3.6	3.8	20.6	21.7	22.1
I	31.5	31.7*	35.1 34.8	175.4	180.2	184.2 182.4
IRL	2.0	1.9	2.1	9.7	9.7	10.7
L	0.55	0.57	0.59 0.57	3.5	3.5	3.6 3.5
NL	10.5	10.4	10.9	62.4	62.5	64.4
UK**	54.3	54.1	55.5 52.9	278.7	264.9	269.1 259.7

* Estimate of peak if no shedding or reduction of load had been applied; real peak was 30.2 GW.

** Statistical figures are not weather corrected, but forecasts are weather-corrected estimates.

Table 3-2:
Peak Demand and Annual Electricity Consumption of EUROPE-10 Countries

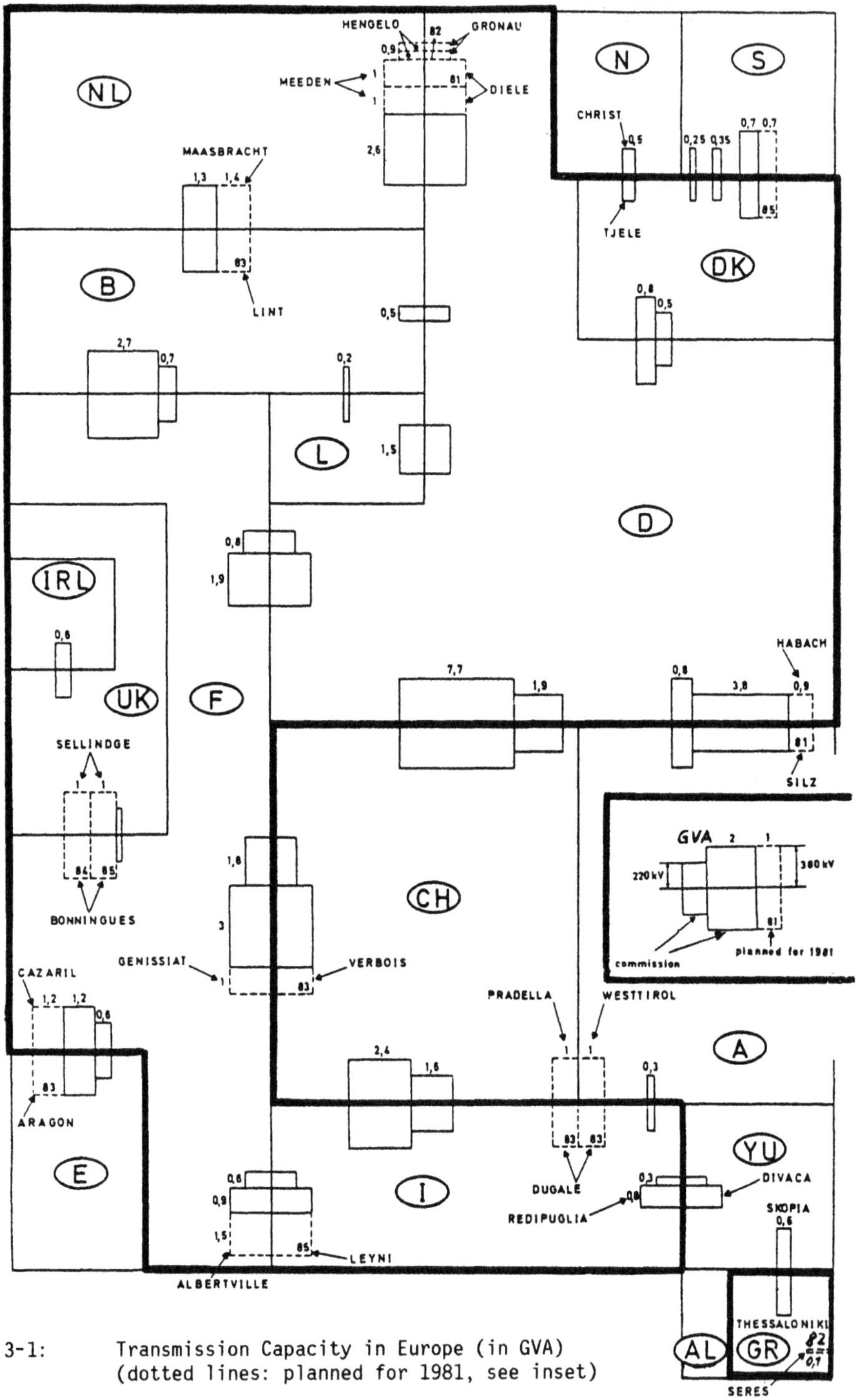

Fig. 3-1: Transmission Capacity in Europe (in GVA)
(dotted lines: planned for 1981, see inset)

3.3 Penetration Problems

The national electricity networks are interconnected with neighbouring countries to ensure great reliability in the supply of electricity. The transmission capacities are illustrated in Fig. 3-1, ranging in the order of 0.5 ÷ 10 GVA. The agreement between the Electricity Boards states that each country must establish its own supply without relying on neighbouring Boards for reserve capacity.

As regards wind energy, the same principle should be followed; i.e. the new installations should be treated according to the national power station mix although the transmission capacity of the networks crossing the border could allow for the power of hundreds to thousands of WECs of the MW size.

As well as these general regulations, some ideas and early system analysis studies of energy exchange have already been published, for instance, to connect the hydro power of Norway to the wind power of Denmark. Although this idea cannot be discussed here, future concepts of wind energy utilization might include special arrangements between neighbouring Electricity Boards to import/export wind power.

3.3.1 Limitation of Wind Energy Utilization by the Existing Transmission Capacity

Single-sited small WECs will be connected to the low voltage system (220/380 V) and will not cause problems under this heading.

For generation capacities of several hundred kW (clusters of small WECs or medium WECs) the 11 kV network is of greater interest as the sites will be located in rural areas or on the outskirts of urban areas and it is in this part of the distribution network that limitations may exist.

In some countries much of the 11 kV network is single-phase overhead line with little scope for reinforcement without considerable expense. Also, in rural areas the load levels are likely to be low on much of the 11 kV network and the network impedance high, which would result in serious voltage control problems with the variable power input from large aerogenerators. For machines of the order of 1 MW, connection at 11 kV is likely to be restricted to the larger 10/50 kV substations or their supply lines. Except at points near transformation from the higher voltage level it is unlikely that more than one or two machines can be accepted onto an 11 kV circuit. Since the density of distribution networks varies enormously throughout the countries it is not possible to give a firm indication of the transmission distances likely to be experienced for connection at 11 kV.

Connection at 33 kV should not present any major problems since being part of the main distribution network it exhibits a much more restricted range of fault levels. It should be possible to accept small groups of machines with total ratings up to 50 MW. Separate generation groups could in many situations be connected into adjacent 33 kV networks, but care would have to be exercised if more than one group is connected to interlinked 33 kV distribution routes. Since the power output from the wind source is variable and not necessarily correlated with load, connection of large amounts of wind generation could aggravate the problem of voltage control and result in more frequent operation of tap changers. Transmission distances to the nearest 33 kV connection could be anywhere in the range 0-20 km in Belgium, Denmark, West Germany, Netherlands and the United Kingdom, or more depending on the load density and the distribution policy of the Area Board concerned.

Where the wind and siting conditions merit the grouping of larger numbers of machines, their total output is likely to be well in excess of the load at the location. There would then be little merit in attempting to connect many machines at the lower voltage levels. The economics of interconnection would probably favour subdividing the remaining machines between a number of separate transformers feeding into the 132 kV system.

Whether the transformers would be grouped together for connection into the 132 kV line would depend on local conditions, since the ease with which transformers could be changed and the limited firm capacity of wind power would not justify much interconnection to improve availability. Total generation capacities up to about 500 MW could be accepted at a major 132 kV supply point.

For large generation capacities exceeding 500 MW the location of the next level, the 275 KV (or even 400 KV system) has to be considered as in most cases transmission of the energy to the consumer centres at a reasonable distance must be assumed.

A more detailed though still general consideration of the situation follows for the various countries.

Belgium/Luxembourg

When the map of wind energy potential (Fig. 2-4) and the high voltage network are compared, it can be seen that the area with high wind potential is mainly supplied by the high voltage transformers located in Eeklo and Izegem with 500 MVA each. In this high wind potential area, the assessment study found approximately 750 suitable sites for turbines of 100 m diameter (3 - 5 MW), equal to a power source of more than 2500 MW. If we assume, as a first order approximation, that 50 % will be directly consumed by the cities in this area, the transmission capacity of the two transformers is close to the surplus power (1250 MW).

The same consideration of the areas of lower wind potential (below 5 m/s) leads to the same finding, i.e. except for short distance reinforcements of the existing network, transmission of the potential power from land-based WECs should not be a major difficulty.

A different result is found for the off-shore potential. Here, the 575 sites in waters of less than 10 m depth would require the installation of separate power lines to transmit the installed power of approximately 2500 MW.

Denmark

Owing to the flat terrain, the wind potential is rather evenly spread all over the country, as shown in Figs. 2-21 and 2-22. The Danish network, illustrated in Appendix 4, is well developed and could be connected to the potential wind power stations without major installation costs.

As the energy potential from land-based WECs already exceeds the national consumption, the off-shore sites within 10 m water depth for approximately 5,300 WECs (equal to more than 4 times the national electricity power) call for very detailed studies to find the minimum costs.

France

Among the sites located in areas with wind speeds exceeding 5 m/s, the coastlines of Languedoc, Vendée, Brittany and Normandy offer a high energy potential.

In the vicinity of Montpellier and Marseilles, 2400 potential large WECs represent a power capacity of 7.5 to 10 GW which - in principle - could be handled in this region by consumption and transmission. The same remark applies to 1500 potential large WECs (4.5 - 7.5 GW) along the coast of Normandy. A limitation due to the transmission capacity might be expected for an intense utilization in the Brittany.

West Germany

The areas of highest wind energy potential (see Fig. 2-16) are supplied by the NWK Electricity Board, as shown in Appendix 4. The density of the high voltage network would enable an easy connection of WEC power stations at relatively short distances.

In 1980, the networks in Northern Germany delivered a peak demand of approximately 10 GW at an installed capacity of 50 % more. By comparison, the total potential of sites for large land-based WECs, as found by this study, equals a peak power of 30 - 50 GW (11000 WECs). The utilization of sites in the high wind zones alone (above $\bar{V}_{10}$ = 5 m/s) would lead to an installed capacity of 7.5 - 10 GW with an annual energy production of approximately 33 TWh.

It may be concluded that, without reinforcement of the coastal network and additional transmission lines to the other German Electricity Boards, the total wind energy potential cannot be utilized but only one of the high wind sites. Inclusion of the potential sites in shallow waters (10 - 20 GW peak power) would further stress this argument.

Greece

The situation in Greece raises a new aspect: the wind energy potential is far greater than the national consumption of electricity (see Table 2-17), hence the transmission network can handle only a small fraction of the total potential. Even if one takes into account only the WEC sites less than 5 km distant from power lines, the potential annual energy would still be too great (by a factor of 7!).

The general arguments developed in the assessment study imply that the network will not be the limiting factor for wind energy utilization. Yet the possibility for such a limit will remain if for some reason special locations with high generating density are selected. Furthermore, it is out of the question to use the high potential offered on the islands by direct consumption and/or transmission.

Ireland

In Ireland, arguments similar to those for Greece can be applied. The high wind energy potential on the one hand and the comparatively low demand on the other enable the application of many siting criteria. Transmission capacity will be only one aspect.

Italy

In Italy, the areas near Genoa in the north and near Naples, Foggia, Brindisi and Rossano in the south seem to offer consistent amounts of potential wind power (about 7.5 and 60 GW respectively). The limited local consumption would imply considerable reinforcement of the distribution and transmission networks, mainly in the south.

The island of Sicily exhibits a vast quantity of sites, but most are in low wind energy regions. It is probable that the local energy consumption can absorb the wind energy production, and no transmission reinforcements should be necessary.

In Sardinia, the grid of which is presently connected to the continent only by a d.c. link of limited transmission capacity (200 MW), the wind energy produced by the great quantity (2500) of sites of high energy level would be limited by the low local consumption.

TheNetherlands

Owing to the dense population of the Netherlands and a well-sized network, no limitation should be expected for utilization of the wind energy potential. This will not exclude the installation of new feeder lines, transformer stations, etc. but the extent of such installations would be comparable to those required for many new generating capacities.

United Kingdom

The comparison of the transmission system and the location of potential sites shows a good fit if dispersed WECs are assumed. A different prospect results by looking at the high energy sites (wind speed above 6 m/s) located in the west of Cornwall, Wales and Scotland. From the transmission point of view, the double circuit lines at 400 kV from Cornwall and Wales are excellently suited to serve as feeders but the problem will come up with the combined utilization of the conventional power stations in these areas.

In Scotland, the transmission lines to the windy zones in the west are far too weak for the utilization of the power available, hence new circuits would have to be installed in the case of large-scale production of energy from wind. Despite some local constraints, the transmission system is unlikely to be the limiting factor in wind energy utilization.

3.3.2 Aspects of the Combined Operation of Conventional and Wind Power Stations

In the preceding section, it was concluded that the existing electricity networks in the countries discussed could transmit only part of the potential wind energy from large WECs. Nevertheless, the magnitude of the identified potential encourages the search for additional findings. As mentioned before, the value of the wind as an energy source will finally depend upon economic criteria. In order to guarantee a constant supply of electricity the interconnection of various types of power stations is obviously required. Hence, the question arises: What are the technically stringent influences for the optimization of interconnection in the near future?

Owing to the specific characteristics of the power station mix and the composition of consumer demand, such an analysis should always be performed for specific networks. Several Electricity Boards have already started this kind of simulation study and three examples are given below (a fourth example is also given but does not originate from a utility).

Example No. 1

The grid management problems of Denmark are determined by two very specific criteria:

- lack of a hydro power station,
- a high degree of cogeneration, i.e. combined heat and power production.

Figures 3-2 and 3-3 illustrate the difficulty of matching the seasonal demands for heat and electricity which limit the possibility of stopping units during low load hours. This leads to difficulties in using wind generated electricity during those hours in winter when the demand for heat is considerable while the demand for electricity is low. The diurnal variation of the electrical load is substantially greater than the corresponding variations of the heat load. The fact that most of the operating units are coal-fired, with a limited control range and high start/stop costs, means that the amount of wind generated electricity which cannot be absorbed by the power system increases rapidly with the installed wind power capacity.

As part of the above-mentioned study, a simulation model has been developed. It simulates the operation of an electrical power system with unpredictable elements of power production from wind turbines. The simulation programme calculates savings in operating costs and fuel expenses as well as the "surplus energy" due to wind production. "Surplus energy" is wind generated energy which cannot be absorbed within the assumed conventional system.

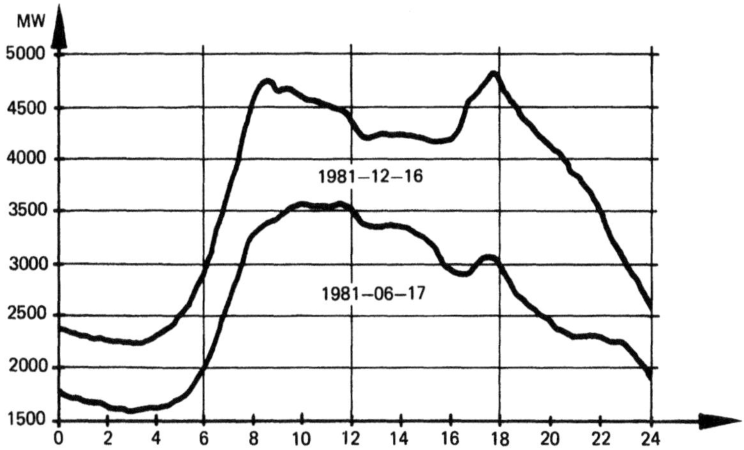

Figure 3-2:
Hourly Electricity Consumption in DENMARK

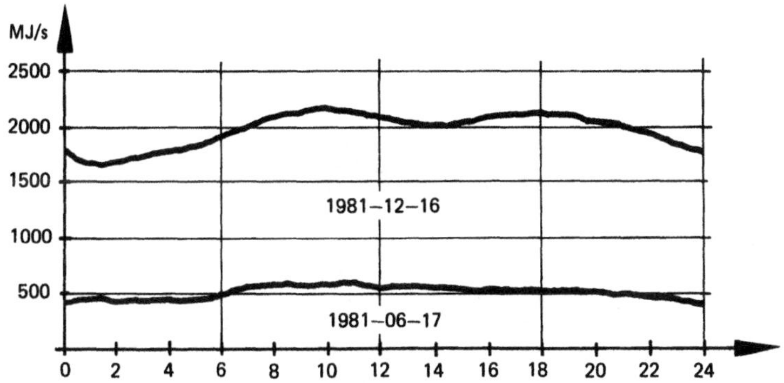

Figure 3-3:
Hourly Heat Consumption in DENMARK

The simulation programme calculates, on a day by day basis, the optimal unit commitment by considering the following factors:

- The electrical load variations during the following two days.
- The heat load variations during the following two days.
- Costs of starting and stopping the available power units (i.e. units which are not unavailable due to forced outage or maintenance).

o Costs of the marginal heat rates of the available power units.

o Agreed exchanges of power with other countries.

The available wind turbines are taken into account in two different ways. In the first case, no wind power unit is included in the unit commitment scheme because of the unpredictable nature of the wind. In the second case, the actual wind energy production is assumed to be known for the following two days.

Having determined the optimal unit commitment, the programme then calculates for each case, hour by hour, the so-called "surplus energy" - this being defined as that part of the actual wind energy production which exceeds the difference between the grid load and the total output power of all conventional production units operated at their minimum capacity. This minimum capacity is predetermined by the heat demand and other constraints which keep the selected units running. In this context the term "forced production" is sometimes used.

The actual wind energy production (production time-series) is based on a power curve with cut-in wind speed = 6 m/s, rated wind speed = 13 m/s and wind data collected at five different geographic locations during a year with normal wind conditions.

Calculations for the Elsam area at the expected 1987 stage give the following results:

Wind energy (Percentage of electricity consumption)	"Surplus energy" (Percentage of wind energy production)	Marginal increase of "surplus energy" (Percentage of increase in wind energy production)
5	12.1	
		> 21
10	16.5	
		> 32
15	21.6	

The amount of "surplus energy" is little affected by forecasting the wind energy production. This is presumably due to the fact that combined heat and power generation units may not be stopped even in cases of considerable "surplus wind energy". It should be stressed that the problem of surplus energy is only an economic and not a technical one. It should, furthermore, be emphasized that the above results are preliminary. Moreover, the results depend very much on all the characteristics of the power system in which the wind turbines are to be integrated.

Another problem is that of matching the power production to the load demand. Coal-fired thermal power stations are slow to respond to changes in electrical load, and the momentary output power of wind turbines is much more unpredictable than the daily electrical load variations, even on a short-term basis.

Comments: In the above-mentioned programme, "surplus energy" is treated as a non-usable "excess energy". It may, however, be assumed that the surplus production will be used for some purpose, for example a heat purpose. The economic value of the "surplus production" is therefore greater than zero. On the other hand, there is no doubt that the specific value of "surplus energy" used in this way will be considerably lower than the normal value of wind generated electricity. In this connection, it is important to conduct investigations to find out how future fossil plants might be modified in order to achieve better cogeneration.

Example No. 2

The arguments from The Netherlands are of interest for two reasons:

Firstly: the power plant interconnection deviates from the Danish one in the use of nuclear power stations and units fired by blast furnce gas.

Secondly: two different opinions are available from the contractors of Action 1 and Action 3, coming from the Netherlands Energy Research Foundation and the Cooperating Electricity Production Companies respectively.

The arguments from the supply side are as follows:

The Dutch electricity supply system consists of a number of regional and municipal electricity companies, linked by a network of transmission lines. The high level of cooperation between the many utilities within the framework of SEP (Cooperating Electricity Production Companies) and SEP's sophisticated interconnection grid give the system a national character.

So long as no storage facilities are available, supply should meet demand at any moment. Hence, an appropriate capacity and type (i.e. response time) of thermal generating plant must be provided and some of it must be kept in reserve to allow for unforeseen changes in demand and unscheduled outages in power plants.

Although variability is an inherent feature of electricity supply, the superposition of a large number of consumers results in a load pattern with predictable average daily and seasonal variations.

The incorporation of a stochastically varying source such as wind power poses entirely new operational problems for the system. Since wind power is not only unpredictable but also unable to adapt to load, the thermal power plant has to follow not only load but also the variations in wind power production, so that wind power can be treated as a "negative load".

An essential problem in the assessment of wind power integration is the limited availability of significant knowledge on temporal variations on time scales relevant to the electricity generation process, spatial variations of wind power, etc.

When compared with the scale of meteorological phenomena, The Netherlands is a small country. Therefore, the behaviour of wind turbine clusters at different locations can be expected to be rather correlated, even for a dispersed siting. This is valid for the time intervals of hours and not of minutes.

Since reliable knowledge of short-term wind fluctuations and the resultant predictability is lacking, SEP, when dealing with the intergration of wind power, provisionally assumed that maximum variations of wind power, considered as a whole, can occur within seconds or minutes, with consequences comparable to those of outages and thermal power plant breakdowns. The presence of this random component in the generation of electricity has already imposed the necessity of back-up generating facilities and of spinning reserve. In total, this represents a rather conservative assumption.

The Dutch electricity supply system is able, even if disconnected from the other UCPTE partners (Union for the Coordination of the Production and Transport of Electric Power), to compensate for the breakdown of the largest generating unit (at this time 650 MW), without unacceptable frequency disturbances. In that situation 650 MW would also be the limit to wind power contribution to the grid if no storage is available.

Another important factor is the time required to restore the initial situation. Here the limiting factor is the ramp rate of on-line plant. The most critical situations occur during rapid load changes, e.g. in the morning. Even if the power level alterations caused by wind power do not exceed the possibilities of primary reaction, the ability of thermal power plant to follow the load variations will determine the potential for large-scale wind power.

In order to illustrate the role of the response capability of thermal plant, the generation system will be considered as a whole. This system consists of a set of power stations characterized by their rated power, fuel, start-up times, range and rate of regulation, etc. Utilities are concerned with operating some of them, both continuously and at their rated powers. This applies especially to combined heat and power and to district heating units, as well as to units fired with blast furnace gas, nuclear and coal-fired plants. Frequent part-load operation of combined heat and power or district heating plants will threaten their efficiency.

In operating a grid contributed to by wind power capacity it would be attempted to optimize the fuel saving due to wind, while part-load operation of blast furnace gas plant might lead to waste of energy when flaring some quantities of available gas. Part-load operation of nuclear and coal-fired plant will also be detrimental to the efficiency of wind power, since the load factor of this type of plant would be affected.

At the present state of wind power development, a conclusive economic analysis of the various factors cannot be made. In order to indicate the effects of the above-mentioned constraints, the situation as expected by 1990 can be examined as an example, assuming that combined heat and power units, district heating capability, nuclear power plant and blast furnace gas fired plant will not be used to follow the load variations. Figure 3-4 shows the expected typical diurnal variations in power demand in the Dutch grid for the peak demand week for 1990, with an expected peak load of 12,000 MW.

The bottom part of demand is met from the following base load power stations:

o industrial combined heat and power plant	1,700 MW
o public utilities combined heat and power plant (district heating[1]), evaporators for water desalination	1,000 MW
o nuclear power plant	500 MW
o blast furnace gas fired plant[2]	500 MW

1) Only the electricity produced during the night in combination with the production of heat in extraction turbines is considered.

2) Only the capacity that must be permanently operated to avoid flaring of blast furnace gas is considered.

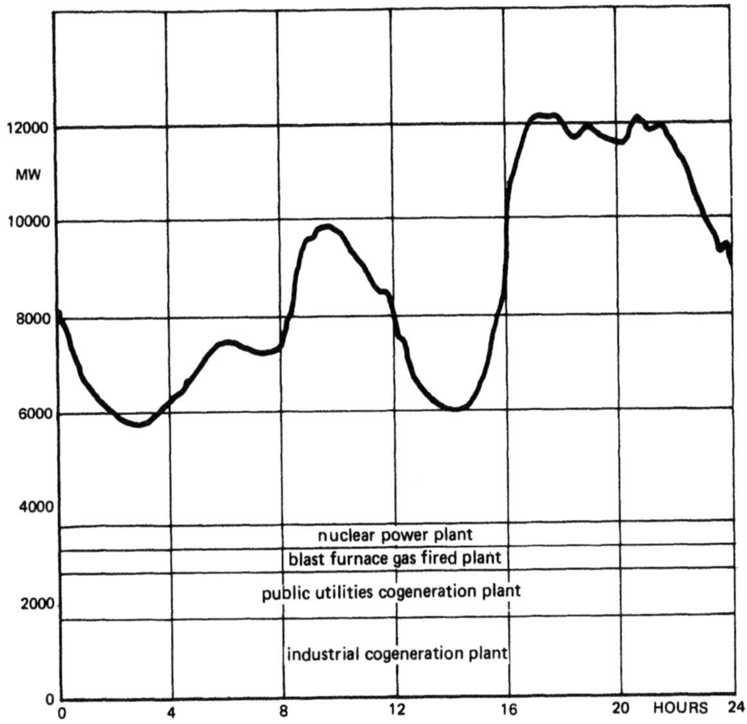

Figure 3-4: Typical diurnal variation in power demand in the Dutch grid as expected in 1990

Further, it should be noted that in the Dutch situation the distinction between intermediate and peak load plant is not sharp. The installed gas turbine capacity is normally used for emergency situations and not for daily peak shaving, as in some other countries.

In view of a certain non-availability of combined heat and power plant, it is obvious that during the lowest load hours at night, only 2500 - 2800 MW can be supplied by wind power and regulable thermal power plant. Only a part of this can be wind power, otherwise no regulable plant will be operating and consequently able to follow the load gradient in the morning and the variations in the output of wind turbine generators (WTGs). This thermal plant has constraints on the rate at which power may be increased or reduced (2 per cent of rated power per minute, within a regulation of 40 per cent and 95 per cent).

The spinning reserve - ready in case of sudden forced outages even of the largest unit in operation - is proportionally spread over the operating units with a load-following duty, so that all units can contribute to the primary reaction.

If no wind power is used, the supply system can safely meet the peak demand of 1990. 6250 - 7000 MW regulable capacity can be operated continuously, during the night at 40 per cent part load. The power output of this capacity may be increased at a rate of 125 to 140 MW/min. Even in periods with high load rises (between 6 a.m. and 8 a.m.) the load gradient is at most 40 MW/min., considerably lower than the following capability of the supply system.

The continuously operated plant (regulable and non-regulable) can meet a load demand of approx. 9100 - 9550 MW. To meet the peak demand some 3100 - 3550 MW plant must be brought on line quickly. They can be regulated at a rate of 60 - 70 MW/min., so that they can meet the peak load safely.

The primary reaction is able to compensate for breakdown of the largest operating unit (at present 650 MW), even if the outage occurs during the most critical hours. To restore the balance a new unit must be started up and brought on line. The disturbance caused by breakdown of a 650-MW unit can be eliminated within 8 min.:

$$\frac{650 \text{ MW}}{(125 - 40) \text{ MW/min}} = 7.7 \text{ min.}$$

During peak hours (ramp rate 60 - 70 MW/min.) it would take approx. 30 min. to restore the original situation. For island operation this would be unacceptable. If the regulable base load plant is run below rated power and some more units are brought on line at appropriate times, a larger spinning reserve will be operational and consequently a shorter restoration time (approx. 10 min.) could be achieved.

The direct integration of wind power poses new operational problems for the system. Assuming that thee will be no loss of wind energy, a smaller amount of base load plant could be operating during the night. With a diminished ability to follow load changes, this thermal plant has also to eliminate the variations in wind power production. By way of example the expected situation in 1990 will be examined, assuming a 1000 MW wind power supply to the grid and supposing that no wind energy losses occur. Since base plant is run to the minimum power output, 1000 MW of wind power will replace 2500 MW of this kind of plant (see Figure 3-5). Consequently only 3750 - 4500 MW of thermal plant which may be regulated can now be continuously operated, instead of 6250 - 7000 MW when no wind power is integrated. The rate at which power may be regulated diminishes also to 75 - 90 MW/min., instead of the initial 125 to 140 MW per minute. To meet the maximum daily demand, 4600 to 5050 MW, units must be started up and regulated. Since during the day "outages" in wind power could also occur, the spinning reserve must now equalize the installed wind power capacity, so that another 1000 MW of thermal plant must be brought on line. If no variations in wind power production occur, it is possible to start up and bring this plant into the range of regulation before the operating capacity has reached its rated power.

Since less thermal plant is operated during the night, the system's ability to follow load variations (including wind power production) will diminish. Even if outages of wind power do not occur suddenly, the primary reaction must also be able to compensate for breakdowns of thermal plant.

In the presence of 1000 MW of wind power, load following constraint is not violated, even in case of sudden forced outages of the largest thermal unit.

Critical situations occur when variations in wind power production superpose important demand changes. If in the morning the wind power production drops by 35 - 45 MW/min. decreasing to nil within approx. 30 min., the power system will just be able to compensate for the resultant load increase and the failing wind power. If wind power drops occur more suddenly, the load-following capability of the thermal system becomes insufficient and loss of load will occur.

Larger amounts of wind power make this situation worse, since a system with a diminished response ability has to eliminate larger wind power fluctuations. Moreover, the situation examined (winter) is not the most critical one, since in autumn and spring the district heating units are still operating at maximum while power demand is much lower.

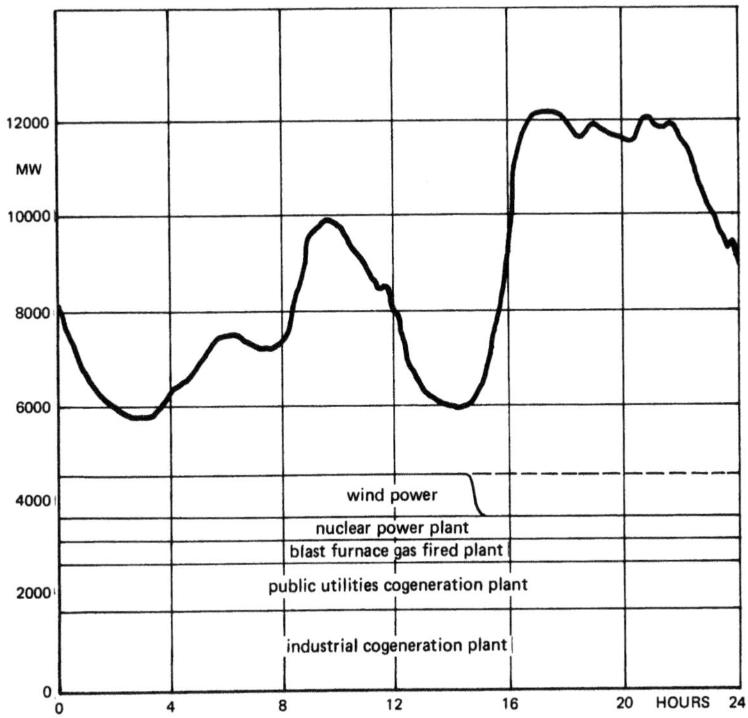

Figure 3-5: A typical diurnal variation in power demand and the new plant mix introducing 1000 MW of wind power

In the absence of relevant information on the major aspects of wind power, the above approach, can provide some indications as to the wind power potential for electricity generation:

- o If further experimental work demonstrates that sudden drops of wind power can occur within a few minutes, the primary reaction capability of the operating thermal plant will be the factor limiting the integration of wind power.
  In this situation the response ability of the operating system will be the prohibitive factor, limiting the integratable wind power to 650 MW (assuming no wind energy losses).
  Taking into account the fact that the power output of the installed WEC capacity is mostly lower than the rated output, this limit may be shifted to - possibly - 1,0 MW, with an adapted spinning reserve.
- o If further research proves that outages of wind power in <u>all</u> situations are not expected to occur within half an hour's time or even longer - either as a result of dispersed siting of wind turbine parks or by a good control of wind power - the primary reaction will cease to be the restrictive factor for the direct integration of wind power.
  Even then, assuming that no wind energy losses occur and, at the same time, no district heating, combined heat and power, nuclear and blast furnace gas-fired units are shut down or

operated below their rated power values, no important amounts of wind power above the 1000 MW mentioned could be integrated by 1990. Larger amounts of wind power reduce the operational flexibility of the power supply system and, consequently, its reliability, since during the night insufficient thermal plant can be kept on line to assure the necessary primary reaction.

A possible extension of this amount depends on the future development of electricity demand and on the share of thermal plant whose output cannot be regulated (e.g. combined heat and power, district heating, nuclear plant) in the plant mix.

o Should it be established that sudden drops in wind power could occur within 30 min. but never quicker than within 3 - 5 min., quick-starting units (gas turbines) could be used to compensate for the variations in wind power production.

Although in this case the response ability of the remainder thermal plant is no longer the restrictive factor, it is obvious that the use of back-up units can only diminish but not solve the problems of direct integration of wind power.

Wind power assisted by back-up facilities cannot follow changes in load demand, but can at best guarantee a constant power output. This constant supply replaces the operation of conventional base load plant, diminishing the number of units which could be operated continuously and during the night at minimum load.

This is in contradiction with the use of a storage system which does not affect the utilization of base load power plant. Since a minimum of regulable power plants have to be taken on line to follow the load in the morning, it is obvious that the integration of, say, 2000 MW of wind power is not realistic. Only the installation of a huge amount of gas turbines (approx. 2.5 times the amount of wind power) would maintain the operational flexibility of the power system. Their high running costs would be detrimental to the efficiency of wind power.

It should be mentioned that the gas turbines from combined heat and power units cannot be used as back-up systems, since no correlation exists between their operation and power demand.

o If further experiments demonstrate that even for a "worst case" the drops in wind power never exceed the ramp rate constraints of the thermal power system, a larger contribution to wind power could be achieved if during the night wind turbine generators are partly operated, or units such as combined heat and power, district heating, blast furnace gas and nuclear plant are run below rated power, resulting in:

- wind energy losses
- and/or diminished fuel savings through combined heat and power generation
- and/or energy losses by flaring blast furnace gas
- and/or a reduced load factor of nuclear plant.

Since the purpose of any operating strategy for a given supply system is to minimize the cost of the electricity produced, in operating a grid containing wind power efforts would be made to optimize its degree of penetration. Accurate wind data and more knowledge of the behaviour of large wind turbine parks, investment cost, O&M costs, operating constraints, etc. are needed to effectuate this complex decision process. It follows that conclusive demonstration results should be obtained as soon as possible.

Example No. 3

The other study in the Netherlands, which started from a statistical treatment of wind power, yielded higher penetration rates because operational problems were not stressed. The basic question for such a study is whether or not the statistics for wind energy differ from statistics for conventional power plants. Selecting the appropriate boundary conditions several authors [3-2] ... [3-5] proved that the applicability of the probabilistic methods used in power plant interconnection can also be used for wind energy integration. Especially the LOLP method (loss of load probability) opened the possibility to investigate the change of the total fuel consumption and the fuel mix with varying strategies of the loading order.

With the aid of the power duration curves, the energy produced the forecast of the electricity demands in 1985 (see Fig. 3-6), and with wind power stations located at Den Helder and Urk, the impact of the wind energy penetration rate was simulated [3-2].

The simplified conventional generating system is shown in Table 3-3. In order to investigate the influence of the chosen unit scheduling strategy, two loading orders were considered. First a very simple scheme was assumed. All available wind power was accepted initially. Then the conventional units were committed in the order shown in Table 3-3, i.e. the order of increasing production costs. All units were allowed to operate at once at their rated capacity when necessary. Calculated results for the strategy, and for varying penetrations of wind power, are shown in Fig. 3-7, variant 1. This figure shows that the total capacity of the nuclear and coal-fired plants is so small that their production is not curtailed by even 4000 MW installed wind capacity. Fuel savings are obtained for oil/gas fired plants and in particular for the smaller and older units with worst fuel economy. One should bear in mind, however, that the calculated fuel savings are too optimistic since no account has been taken of any possible operating constraint of load-following penalty.

In the second loading order a more elaborate strategy of plant operation was assumed. First, nuclear plant was considered inflexible, operating at rated power whenever available. Secondly, since start and stop procedures for the large fossil-fired plants are undesirable from a technical and economic point of view, these units ($\geqq$ 375 MW) were kept on-line even if available wind power would have allowed the shut-down of some of these units.

Further, a programme operation of the gas/oil-fired plants was applied, in which the power levels of these units were changed stepwise: They were all first scheduled to operate at their technical minimum (30 - 40 % of rated power) to serve low system demand, and next allowed to increase their output for load-following. Finally, under normal load-following conditions the fossil-fired units were not allowed to operate at their rated capacity, but only at about 90 % of this value; the last 10 % was used only for system peak demand and in emergency situations. In this way a reasonable modelling of spinning reserve could be established. The loading order and programmemed operation of the plants are shown in Table 3-4. Results for this strategy are given in Fig. 3-7, variant 2.

SEP (Cooperating Electricity Generating Companies) have reservations about these findings because the studies assumed that by 1985 only 1200 MW coal plant will be operating, whereas about 1900 MW coal-fired units are already in operation and another 1650 MW is being converted from oil to coal-firing. Also no allowance was made for about 1100 MW cogeneration plant operating by 1985 (today about 800 MW). This type of plant cannot have a load-following function. Both these assumptions lead to an unacceptable overestimate of the power system flexibility. Moreover, no allowance was made for load variations with time of year, the situation examined (winter) being the least critical one.

Table 3-3: Simplified Scheme of the Power System in 1985.

Fuel	Size (MW)	Number of Plants	Installed capacity (MW)	Forced outage rate (%)	Maintenance unavailability (weeks/year)	Age*
uranium	450	1	450	12	6	a
uranium	50	1	50	12	6	b
coal	600	2	1200	10	5	a
oil/gas	575	6	3450	10	4½	a
oil/gas	375	10	3750	12	4	a
oil/gas	200	5	1000	12	4	a
oil/gas	125	3	375	12	4	a
oil/gas	25	10	250	12	2	a
oil/gas	200	5	1000	12	4	b
oil/gas	125	9	1125	12	4	b
oil/gas	25	5	125	12	2	b
total			12775 MW			

* Age:

a ≦ 15 years

b > 15 years, < 20 years

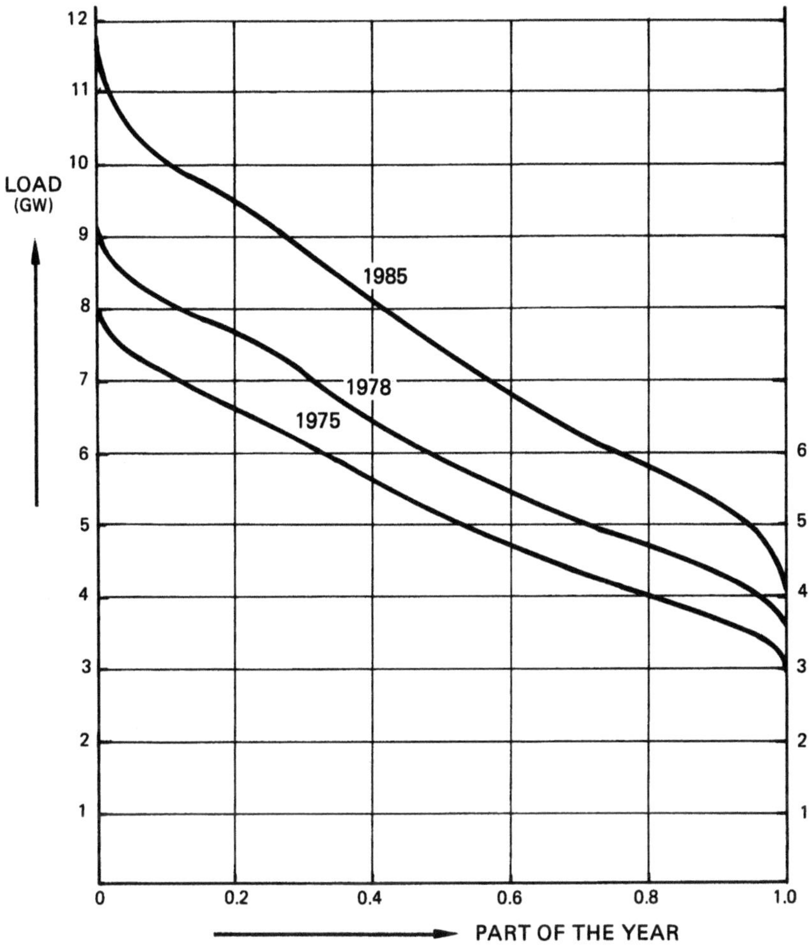

Fig. 3-6:
Load duration curves of electricity demand for the years 1975 and 1978 (actual values) and for 1985 (extrapolated values, including forecast uncertainty).

Table 3-4: Load ordering of the conventional power plants and wind potential (variant 2)

Fuel	Size (MW)	Number of Plants	Load in Percentage of Rated Power
uranium	450	1	100
uranium	50	1	100
coal	600	2	90
oil/gas	575 - 375	16	30
wind	variable		
oil/gas	200 - 125a)*	8	40
oil/gas	200 - 125b)**	14	40
oil/gas	575 - 125a)	24	90
oil/gas	200 - 125b)	14	90
oil/gas	25a)	10	100
coal	600	2	100
oil/gas	575 - 125a)	24	100
oil/gas	200 - 25b)	19	100

* a) plants not older than 15 years
** b) plants older than 15 years.

Summarizing both Dutch studies, the immediate possible penetration rate can be seen to be in the order of 5 % to 10 % of the conventional capacity whilst improved operational strategies might in the order of 20 % or even 30 %. As no technical limitation was cited, the choice of an optimal stragegy depends on economics, and since rather conservative assumptions were used, the chances for a justified higher than 10 % penetration rate are good.

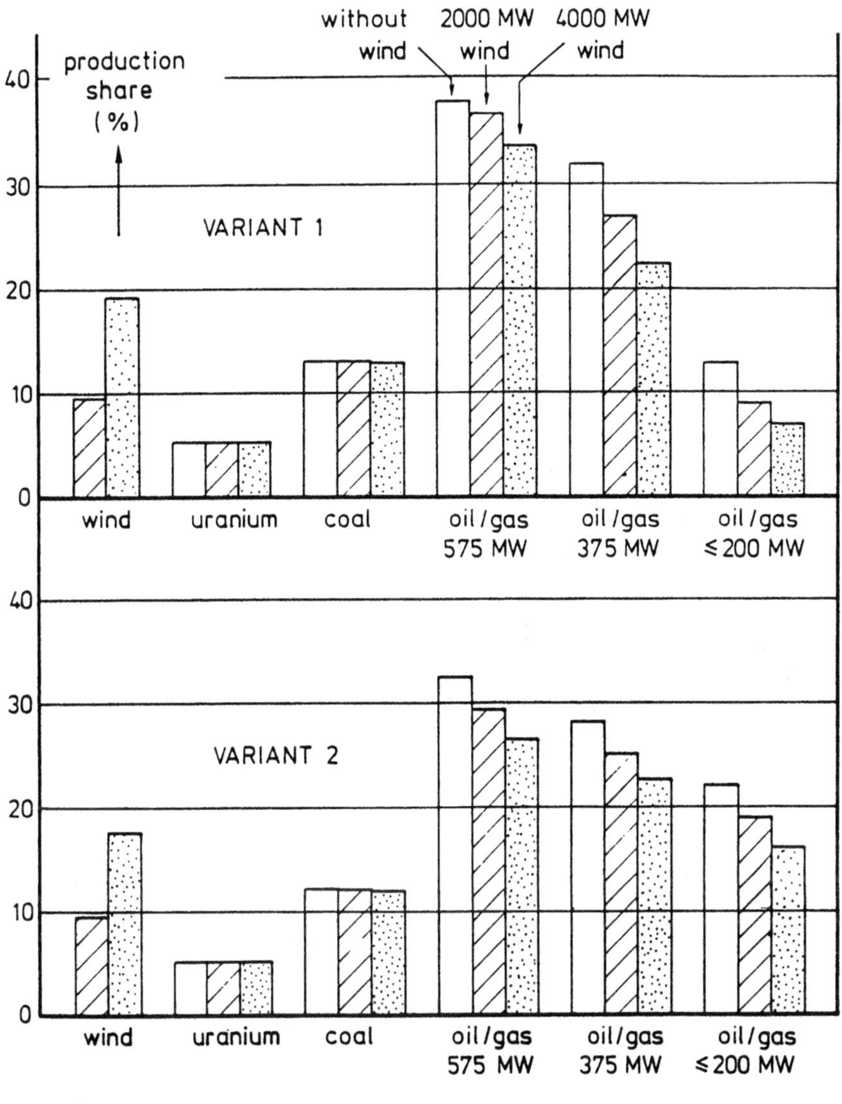

Fig. 3-7:
Production shares of the different plant types for various penetration levels of wind power and different loading or strategies.

Loading order:

Variant 1:	wind first, then conventional plants in the sequence of Table 3-1
Variant 2:	sequence of Table 3-2.

Example No. 4

Another study, performed by CEGB (the Central Electricity Generating Board) for the United Kingdom, should be cited as it incorporates a large amount of nuclear energy. The wind data for 1974 and the national demand curves for 1974 were used for each hour of the year.

The philosophy adopted for this study was that wind energy would be used as a means of saving fossil fuel, and that it would not be economic for wind energy to displace nuclear energy. This allowed a simple system simulation to be used with a plant mix comprising nuclear and large fossil-fired plant. The simulation was organized to study the limitations imposed by limited flexibility of fossil-fired plant and to examine the effect of different proportions of nuclear plant.

The proportion of nuclear plant capacity in relation to the average cold spell (ACS) demand could be varied in the simulation. It was assumed that the nuclear plant would operate at maximum output except when required to part-load in order to match reduced consumer demands: however, the programme did not part-load the nuclear plant to accept wind energy. This limits the load cycling on the nuclear plant but is a pessimistic assumption in respect of the wind power. No allowance was made for the effect of pumped storage on the operation of the system. In practice, flexible operation of storage plant would improve the case for wind power.

Output from installed wind generation is accommodated on the system by part-loading the fossil-fired plant which would be normally required on the system to meet the consumer demand. At the present time the minimum part-load capacity of the large fossil-fired plant varies from unit to unit; however, it is expected that with modifications all the large fossil-fired plant could operate at a minimum continuous load of 50 % maximum continuous rating (MCR). This value is used in the study. For comparative purposes only, a minimum load of 20 % MCR has also been examined. This load is only likely to be achieved on a small number of the existing fossil-fired units, but could be achieved on units specially designed for this operating regime if this were shown to be worth while. The 20 % load also implies costly oil consumption.

Purely for comparative purposes, simulations were also made for a completely flexible system with all the thermal plant able to part-load to 50 % MCR.

Each of the four sites for wind power plants was assumed to have the same basic installed capacity. Then, using the wind-turbine-generator characteristic and the recorded wind speed, the electrical output from each site for each hour of the year was calculated and summed for the 4 sites. This assumes that the arrays will have 100 % availability, and can develop rated power output regardless of the direction of the wind.

It was assumed that, at any time, the combined fossil and nuclear plant was equal to the consumer demand. A fixed amount, defined as a proportion of the maximum yearly demand during average cold spell (ACS) conditions, would be nuclear plant.

The system and plant performance data were used to calculate the amount of wind power that could be accepted for each hour of the year. This was compared with the available wind energy and hence the wind energy utilization found. Cumulative totals of the available wind energy, wind energy utilized, and fossil energy savings were calculated.

Results. Three factors were used to compare the results:

a) Wind Energy Utilization Factor

$$K_W = \frac{\text{wind energy utilized}}{\text{wind energy available}}$$

b) Fossil Energy Saving Factor

$$K_F = \frac{\text{fossil energy saving}}{\text{fossil energy required without WTGs}}$$

c) System Energy Saving Factor

$$K_S = \frac{\text{thermal energy saving}}{\text{total energy required by system}}$$

Figure 3-8 shows the proportion of wind energy utilized for 10 % and 30 % nuclear planting ratios. The minimum load on the fossil-fired plant is 50 % MCR, and the wind-turbine-generator has a characteristic of $V_R = 1.5\ \bar{V}$. In the case of 10 % nuclear plant, typical of present UK system conditions, it can be seen that 20 % wind power could be accommodated on the system with only 2 % of the wind energy being lost. With 30 % nuclear plant the wind energy lost would increase to 15 % if it is assumed that nuclear plant is not part-loaded. The percentage of fossil energy saving with 20 % wind power capacity and 10 % nuclear plant is about 17 % of the fossil energy. With 30 % nuclear plant, the percentage fossil energy saving is increased but in absolute terms the saving is less since the annual fossil energy requirement is lower.

Figure 3-9 compares the wind utilization factors when all the thermal plant is part-loaded to accept wind energy onto the system. 20 % and 50 % MCR minimum loads are compared. Part-loading the nuclear and fossil-fired plant to 50 % MCR allows the wind power capacity to be increased to 30 % with only 2 % of the energy lost. The unlikely situation of 20 % part-loading on all thermal plant would allow very large amounts of wind power to be installed on the system before the energy loss becomes significant.

Figure 3-10 shows the effect on wind utilization of different wind-turbine-generator characteristics. The characteristic of $V_R = 2.3\ \bar{V}$ gives a better utilization factor than the $V_R = 1.5\ \bar{V}$ design. This is because the wind energy available for use is much less with this design and hence system restrictions do not limit the utilization until there is a larger installed capacity. The thermal energy saving with the $V_R = 2.3\ \bar{V}$ design, is however, much less than the 1.5 $\bar{V}$ design, because the higher wind speeds occur less frequently and there is more time at zero output.

A more complete study model has been developed by Epsom, allowing a range of renewable energy sources to be examined in order to assess the potential impact of different plant mixes on system operating constraints and economics.

Limitations. The study shows that wind power up to about 20 % of the ACS level can be accepted onto the system for the present level of nuclear planting without a significant reduction in the value of the wind energy. However, a number of the assumptions made in the study would affect this estimate.

No allowance was made for variation of plant availability with time of year and while this would have only a limited effect on the wind power capacity because of the multiplicity of small units it would affect the nuclear plant. A lower availability of nuclear plant in summer to allow for annual maintenance would marginally increase the amount of wind power which could be accepted without reduced economic benefit.

The inclusion of pumped storage on the system would also increase the total wind energy which could be economically accepted. However, the economics of storage plant are such that they must be assessed as part of a total system operation package and not just as a means of improving wind energy utilization.

Part-loading fossil plant to accept wind power incurs additional operating costs compared to two-shifting, owing to the lower efficiency of part-loaded plant and to out-of-merit operation, if the plant is part-loaded for longer than the economic shutdown time. These additional costs would be equal to the no-load heat cost, or about 15 % of full load cost at the most. In practice, some measure of predictability should permit part of the total wind capacity to be accepted by shut-down of fossil plant and this aspect requires more detailed study.

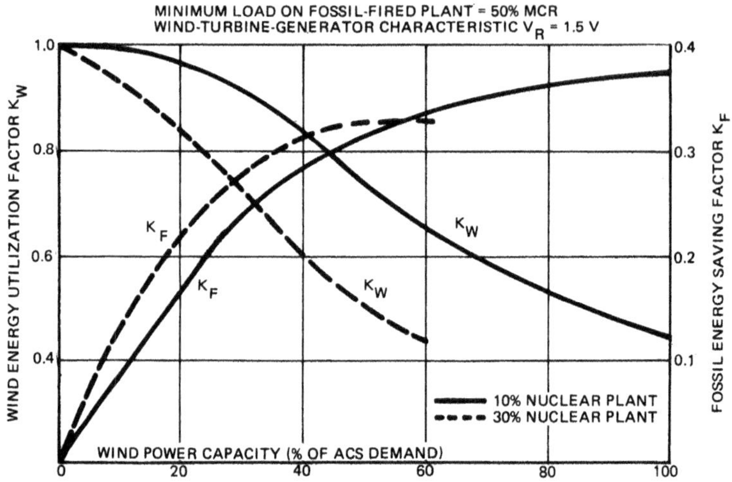

Fig. 3-8: Wind Energy Utilization and Fossil Energy Saving Factors

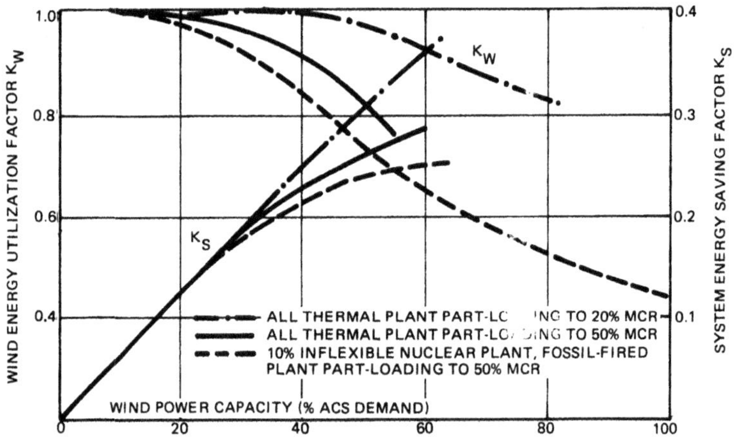

Fig. 3-9: Comparison of Changes in System Regulation Capacity

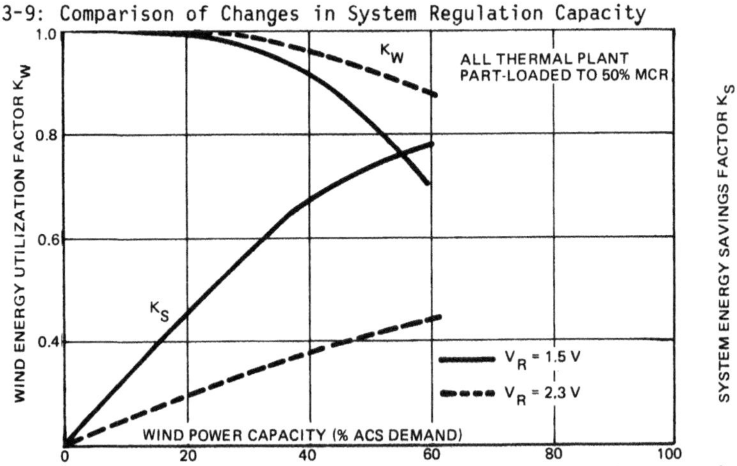

Fig. 3-10: Comparison of Different Wind-Turbine-Generator Characteristics

The data used are from 10 m height for inland and coastal sites where the diurnal variations in wind speed show that the highest wind speeds occur in the afternoon. However, the limited data now available from off-shore sites show that the highest wind speeds occur at night. In both locations the diurnal effects at hub height may well be different from those at 10 m height and this needs further detail study. No allowance has been made in the study for these differences in wind characteristic. This may be important since the system regulation capacity available to accept wind energy onto the system varies throughout the day, and is at its lowest value overnight when the output from off-shore arrays will tend to be at its highest value. In this respect the study data would give optimistic values for wind utilization, but this would only be significant for large installed wind power capacities.

An important aspect of the machine characteristic from the system point of view is the way in which the array output changes as the furling speed is approached. In the simulation the array output is reduced to zero instantaneously at the furling speed. This would not be satisfactory in a large array of say 1000 - 2000 MW maximum output. In practice, as the furling speed is approached a large array must be shut down in a controlled manner to match the rate of load increase achieveable on the part-loaded thermal plant. Array shutdown over a 10 - 15 minute period should be adequate.

Since wind power is inherently variable, its variation will be additional to the system load variation and in certain circumstances short-term variations may impose additional constraints on the operation of other plant, necessitating additional spinning reserve or increased rejection of wind power capacity. This is only likely to be a significant factor for large installed wind power capacities but does merit further study.

3.4 Conclusions

The results of the four simulation studies may be summarized as follows:

- No. 1:
  WEC installations will increase the temporary amount of surplus electricity owing to the influences of co-generation (heat and power). If the WEC plants are capable of supplying 5 % (15 %) of the national electricity consumption, the useful electricity would only be 4.4 % (12 %).
- No. 2:
  Without major modifications, the power plant interconnection chosen could accept in the order of 8 % of the peak demand from wind energy plants.
- No. 3:
  The same plant interconnection could utilize WEC plants from 15 % to 30 % of the installed conventional capacity if elaborate operational strategies are applied.
- No. 4:
  In a power station system with 10 % nuclear plants it can be seen that 20 % wind power could be accommodated on the system with only 2 % of the wind energy being lost if the fossil plants were part-loaded. With 30 % nuclear plants the wind energy lost would increase to 15 % if it is assumed that nuclear plant is not part-loaded. The study assumes fossil stations part-loaded down to 50 % of MCR, which implies modifications compared to the existing plants.

These findings are in agreement with the studies made in Germany by Jarass [3-4] and in Italy by D'Amico [3-6]. Interconnection of wind power plants and conventional plants would be feasible. The degree of penetration depends on the details of the specific grid system, but for WEC plants, a share of 10 % to 20 % of the installed conventional capacity seems to be realistic. This penetration rate will not cause transmission problems for the grids, hence the interconnection and the onset of economic penalties will be the limiting factors for the utilization of wind energy.

For the reasons given above, the potential evaluated in Chapter 2 will be drastically reduced. To deduce the order of magnitude of acceptable wind power, the expected power plant capacity for Europe-10 was taken from the UNIPEDE report [3-1] (400,000 MW for 1986) and reduced by 50 % of the French and German values owing to the location of wind plants in the coastal grids only. The resulting plant capacity of approximately 300,000 MW will lead to a tolerable capacity of WEC plants in the order of 30,000 MW to 60,000 MW, corresponding to 10 % - 20 % of the total capacity, generated by 10,000 to 20,000 units of 3 MW size, of 5,000 to 10,000 units of MW (100 m diameter). This installed capacity will be sufficient for a share in electricity production of approximately 10 %, as shown in Table 2-5. It must be pointed out again that these findings for Europe as a whole are based on large WECs.

There will be a certain contribution by small WECs owned by auto-producers. As the correct capacity credit is still unclear and a matter for future studies, no credit was included in these figures.

These figures may appear more significant when the 1980 plant capacities are recalled: 33,000 MW for nuclear power stations and 59,000 MW for oil-fired thermal monovalent plants. It should be noted that this comparison is not intended to indicate the type of replacement.

To summarize, investigation of limiting factors for the utilization of wind energy has demonstrated:

- o limits set by the national consumption of electricity,
- o up to this consumption level, no limitation by the transmission system, with the exception of a few special areas,
- o no real technical but probably economic reasons for the limitation of combined operation of conventional and wind power stations.

Bearing these limitations in mind, wind energy can be regarded as a real alternative - or better, additional source of electric energy.

4. STATE OF THE ART OF WIND ENERGY UTILIZATION

The preceding chapters have identified the availability of wind energy and sites, as well as the possibility of transmitting the energy and of combined operation of the various power plants. Some indications were given that a certain penalty in production efficiency might be expected if the penetration rate exceeds 10 % - 30 % of the conventional power park. As such a limitation will finally be based on economic criteria, the next step within the assessment study is to look for technological constraints before analysing the economic aspects.

The requirements for connecting small or large WECs to the grid are often neglected but these requirements will be essential factors in the costs and reliability of wind generation. The grid interface problems will therefore be discussed at greater length. Another aspect will be the European market, coupled with the need for future standardization.

4.1 Technical Maturity of Wind Energy Converters

The Final Report of Action 4 will ciscuss the details of the existing WECs and their technical state of the art. Therefore only a short presentation of some characteristic features will be included here.

4.1.1 Technology of Wind Energy Converters

The theory of extracting energy from an airstream is based on the analysis by Betz (1926), resulting in the following equation for the maximum power:

$$P_{max} = 0.593 \frac{\delta V_o^3}{2} A_R$$

where	P	=	power	in [W] = [Nm/s]
				= [kgm²/s³]
	δ	=	ambient air density	in kg/m³
	V_o	=	undisturbed wind velocity	in $\frac{m}{s}$
	A_R	=	rotor swept area	in m²

The "Betz coefficient" is limited to C_{Pmax} = 0.593 as any higher loss of energy would cause stagnation of the air flow.

But this highest conversion efficiency will be reduced if low tip speed ratios (tip speed divided by wind speed) are used.

Furthermore, the drag coefficient of the rotor blades and the solidity of the rotor disc - in practice the number of rotor blades - will also cause losses, as illustrated in Fig. 4-0. Most of the design philosophies can be discussed with the aid of this Figure.

Obviously, high tip speed ratios offer high efficiencies, therefore modern wind turbines tend to be built as "fast runners". But the aerodynamic quality must be high, too, otherwise the drag losses will outweigh the gain by high tip speed ratio. Profiles with lift-to-drag ratios up to 200 have been developed but the very clean surfaces will show a certain roughness after some operational time, causing a loss of up to 50 % in the L/D ratio. For this reason, many manufacturers optimize between aerodynamic quality and rotor solidity because a lower L/D ratio can be compensated by a higher number of blades and most of the small WECs are being produced with 3 or even more blades.

In practice, well designed machines will show efficiencies in the vicinity of 45 % at the design point, but many machines do well with lower efficiency, as the energy production costs depend not only on high efficiency but also on low machine costs and maintenance.

During the centuries of wind utilization, many designs have been proposed. By far the greatest experience is associated with the propeller type with horizontal axis, and the WECs installed in great quantities up to now are of this type. Another system, the Darrieus type with vertical axis, promises to become compatible but as only a few machines have been developed, the final answer must be postponed.

The horizontal axis machines are built with two different rotor positions: in front, or aft of the tower with respect to the wind direction. There are good technical arguments for both solutions, the main one for the upstream position being the reduction of fatigue loads caused by the blades passing through the tower shadow.

The load assumptions for the design of WECs have to consider two different classes: static loads and dynamic loads. As the lifetime of a WEC should exceed 100,000 to 200,000 hours of operation (about 300 million revolutions) almost all components must be designed in accordance with fatigue principles. This consideration applies especially to the rotor.

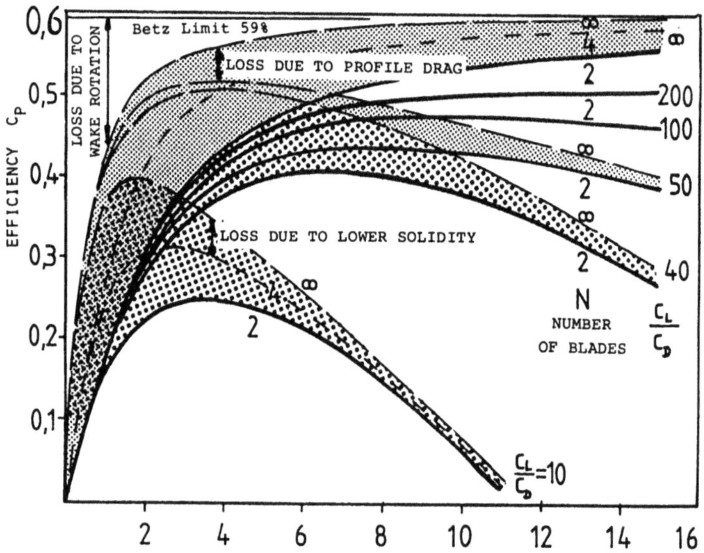

Figure 4-0:
Main Aerodynamic Influences on the Efficiency of Wind Turbines

The WECs of interest to this assessment study are designed to produce electricity. The torque of the rotor drives a generator after the appropriate speed increase by a gearbox. For simplicity, synchronous or induction generators are being used; they will be discussed in Section 4.2 in greater detail. As the frequency of grid-connected machines must be maintained within very narrow tolerances, the speed control of a WEC needs great attention.

As a general rule, two different solutions are applied:

1. The size of the induction generator is such that the rotor speed is controlled by the strong connection to the grid and, in the case of high wind speeds, the aerodynamic stall prevents overspeeding. The main advantage is a simple design of the rotor, as fixed blades can be used.
2. The rotor speed is measured, and the control subsystem is activated to change the pitch angle of the blades. This "active control" requires expensive components in the mechanical, hydraulic and electronic controls.

For the time being, the fixed-blades solution seems to be restricted to rotor sizes below 100 kW because many different operational conditions must be provided for, such as start-stop procedures, redundancy in cases of emergency, the parked position at extreme wind speeds, etc.

A rather new technology, the AC-DC-AC inverter, is being introduced to wind energy converters as it enables a very accurate conditioning of the energy delivered to the grid without the necessity of fixing the rotor speed. In the following chapters, this device will be referred to as "electric control" and there are good reasons to expect system improvements by the future application of this technology.

Wind energy technology has already matured to a special branch within the engineering sciences and many textbooks present detailed studies. Within the scope of this report, the progress of the state of application is of greater importance than specific engineering topics. Therefore the quantity of converters installed will serve as a measure of the experience acquired.

An American report (V. Nelsen, W. Texas State University's Alternative Energy Institute) claims a total of 6,300 units installed from 1970 to the end of 1982 by 37 manufacturers. 14 of them serve as battery chargers (below 1 kW) but one company delivered over 1,000 1.8 to 40-kW machines and another company 260 50-kW turbines. For 1983, a total production of 10,500 units equal to 223 MW was planned.

By November 1982, a total of 737 commercial aerogenerators had been installed in Denmark with the following sizes:

rotor diameter	≦ 8 m:	252 turbines,
rotor diameter	8 - 15 m:	356 turbines,
rotor diameter	< 15 m:	129 turbines.

Worldwide, the quantity of small WECs connected to a grid might well be in the order of 10,000 units. As many of them have been installed under boom conditions recently, the question of reliability and operational experience cannot yet be answered.

For small WECs, operational results for many units are already available from Denmark. The analysis of these WECs after the November gales in 1981 (wind speeds exceeding 35 m/s in 10 min average) revealed [4-1]:

- o total operational time accumulated: 2,000,000 h
- o 9 of 250 windrose WECs suffered total collapse (approx 4 %) and 30 % substantial damage;
- o out of 500 units of different design, only 5 % experienced serious failure and 2 % collapsed.

As the windrose is a multiblade type and WECs for electricity production will be designed with 2 or 3 blades only, the last category of turbines is of interest for further analysis.

The failure types of these 500 units were as follows:

yaw control failure	2.11 %
hub failure	0.88 %
blade tip failure	0.71 %
bearing and gearbox failure	0.59 %
blade buckling	0.35 %
blade stay failure	0.18 %
blade root failure	0.18 %
total	5.00 %

The reasons for these failures were:

failure due to static loads	28 %
failure due to fatigue loads	21 %
insufficient power regulation	21 %
underdimensioned mechanical/ aerodynamic brakes	11 %
miscellaneous	19 %
	100 %

It should be noted that this analysis covered commercial units as well as privately built ones.

Another statistical evaluation of one company [4-2] dealt with two WEC types (30 and 55 kW rated power) in a total of 32 installations after 120,000 operational hours. The machines produced approximately 500 kWh/m^2 year and operated for approximately 5300 h/year. The total service experience showed that operational expenses (repair, maintenance etc.) can be kept below 1 - 2 % of the investment per year.

A noticeable tendency is that fewer home-made WECs are being installed and the number of companies participating in the production of WECs is increasing.

For large WECs, the greatest operational experience has been gained in the USA. The four experimental units Mod-OA located at several sites accumulated a total of 35.000 operational hours by summer 1982. Of course, failures were experienced, but the results gave the confidence necessary for further development of larger machines. The three Mod-2 units (2.5 MW; 90 m dia.) installed at Goldendale, Wash. in 1980/81 suffered some control problems but accumulated more than 3,700 operational hours by autumn 1982. The results encouraged the U.S. agency to start the programme for the development of 5/7 MW machines.

Although a continuous potential for further improvements can be seen for the small WECs, they have already achieved a first maturity level and have therefore become available on the basis of series production with batches in the order of approximately 50 to 100 units.

The large rotors range from the prototype stage close to small quantity production. They encounter more design and manufacturing difficulties than the small units, but on the other hand, they are developed with a much higher level of engineering thoroughness. A total of 11 units with more than 1 MW power have been installed in the world up to now.

4.1.2 Analysis of Commercially Available WECs

The large quantity of operational units cited in the preceding section raises the question of a general classification of some characteristic values. In view of the vast range of WECs already developed, and due to their different test and operating conditions, it must be accepted that the results may differ considerably. When looking for machines with a complete set of data, the WECs type-tested by the Danish Risö Test Centre, representing more than 750 operational small WECs, eased the process of classification very much and therefore the Danish designs serve as a backbone for the analysis. Moreover, the Danish wind classification helped to put some order into the jungle of specifications. Nevertheless machines from other countries also fit well into the overall consideration.

Table 4-1 was developed by adding more and more converter types in several loops. The table represents more than 50 different designs ranging from 4 kW to 4,000 kW. The selection of the turbines included reflected only the availability of complete data sets. In many cases, almost identical values were found for different manufacturers.

Great care was taken to consider only types which are available on a commercial basis or are being developed for series production. The borderline between the two categories falls between the 265 kW and 630 kW-units in the table.

The columns of the table are self-explanatory except the definition of wind speeds. While the rated power (column 4) is associated with the actual wind speed at hub height, as described in the Betz equation, the annual energy production is associated with the mean annual wind speed at hub height. So long as the same annual wind distribution (see Fig. 2-1) is applied, comparisons are justified. But the degree of accuracy of the averaging process and the different climatic conditions in North and South Europe imply that the mean annual wind speed data should be taken as a first-order approach in this respect. Bearing this reservation in mind, a systematic order was arrived at which may serve as a reasonable approximation for the time being.

To start with the analysis, Fig. 4-1 relates rated power to rotor diameter and rotor swept area, covering 4 orders of magnitude in power. Although the technology seems to be well suited to small and large WECs, the plot exhibits the non-existence of a strict design rule for selecting the generator size. Differences of 50 % in maximum power rating can be observed for the same rotor diameter, indicating a design for a high wind speed location or simply oversizing of the generator. A comparison of different WECs on the basis of rated power only is not advisable.

The next diagram (Fig. 4-2) supports this argument. The specific rated power per swept area increases with rotor diameter, mainly due to higher λ values (tip speed divided by wind speed) and the improved aerodynamic quality of the blades. Yet the actual machines show optional influences for the maximum power. For instance, the 2 MW design of the Swedish 75 m rotor has a power rating of 450 W/m^2 in the prototype version but could be uprated to about 700 W/m^2. The American Mod-2 (90 m diam.) has a layout for 390 W/m^2 for the prototypes installed. The reasons behind these findings stem from the development strategy. A lower power rating reduces the fatigue problems - but also the annual energy production, as will be discussed in the next figure. Hence an economic optimization process will finally guide the selection of the generator size.

Interpretation of the specific annual energy is facilitated if the Danish classification of the wind zones (s. chapter 2) is used, as shown in Fig. 4-3. The strong dependence of rotor size stems mainly from the tower height, as the wind potential increases with height. This fact explains the levelling-off of the bands at low values, the very small units being mounted on towers proportionally higher than those of the large WECs. The influence of the power rating is no longer pronounced, though still noticeable - the reason being the very short intervals of high wind speed availability.

Table 4-1:
Comparison of Prices and Energy Production of Wind Turbines

MANUFACTURER	1 ROTOR DIA-METER	2 ROTOR SWEPT AREA	3 HUB HEIGHT	4 RATED POWER	5 SPECIFIC RATED POWER	6 ANNUAL ENERGY PRODUCTION	7 SPECIFIC ANNUAL ENERGY PRODUCTION	8 PRICE INCL. INSTAL.	9 PRICE PER SWEPT AREA	10 PRICE PER RATED POWER	11 INVESTMENT PER ANNUAL ENERGY(8:6)	12 EQUIV. UTILIZATION PERIOD (6):(4)
	m	m^2	m	kW m/s	W/m^2	MWh/a $\bar{v}$[m/s]	kWh/m^2a	10^6ECU	ECU/m^2	ECU/kW	ECU/kWh	h/a
Static Advice T 24	5.4	23	12	4/16	174	16 / 7.1 11 / 5.8 8 / 4.9 4 / 3.8	700 480 350 170	0.008	360	2078	0.52 0.76/20 a 1.04 2.08	4000 2750 2000 1000
Polenko WPS 8	8.5	56.7	20	15/13	260	43 / 7 31 / 6 21 / 5	760 550 360	0.022	390	1473	0.51 0.71/20 a 1.05	2867 2066 1400
Kuriant KE 15 K	11	95	18	18.5/12.5	195	64 / 7.4 43 / 6.1 31 / 5.4 16 / 4.3	674 452 326 168	0.021	218	1100	0.31 0.48/20 a 0.65 1.28	3460 2320 1675 860
AEROMAN	11.6	106	10	20/11.1	200	60 / 6.8	570	0.047	470	2340	0.78/20 a	3000
Bonus and Vestas	15	177	18	55/16	310	174 / 7.4 116 / 6.1 87 / 5.4 46 / 4.3	983 655 490 260	0.043	240	774	0.25 0.36 0.49/20 a 0.92	3160 2110 1580 836
Polenko WPS 16	16	201	24	60/13	300	132 / 6.5 111 / 6 90 / 5.5	660 550 450	0.052	260	873	0.40 0.47/20 a 0.58	2200 1850 1500
Polenko WPS 18	18	254	30	100/13	390	240 / 7.5 180 / 6.5 150 / 6 108 / 5	940 710 590 425	0.084	330	842	0.35/20 a 0.47 0.56 0.78	2400 1800 1500 1080
Aerowatt 100 KFT 13	18.2	260	30	100/13.8	385	250 / 7.5	810	0.221	851	2210	0.88/20 a	2500
Windane 29	29.4	674	29	265/12.5	390	630 / 6.6 485 / 5.8 290 / 4.9	935 720 430	0.552	820	2078	0.87 1.13/20 a 1.90	2380 1830 1090
Nibe B	40	1257	40	630/13	500	1440 / 7.5	1143	1.11	880	1770	0.77/20 a	2280
AEOLUS	75	4418	78	2000/12.8 3000/14	450 680	6700 / 7.8 7700 / 7.8	1516 1740	4.75	1075	2380	0.71/30 a 0.62	3350 2570
WTS 3 WTS 4	78	4778	80	3000/14 4000/15.1	630 840	8000 / 7.8 8300 / 7.2	1670 1740	4.75 7.75	992 1620	1580 1940	0.60/30 a 0.93	2670 2075
Mod-2	90	6362	60	2500/13	390	8000 / 7	1260	7.45	1170	2980	0.93/30 a	3200

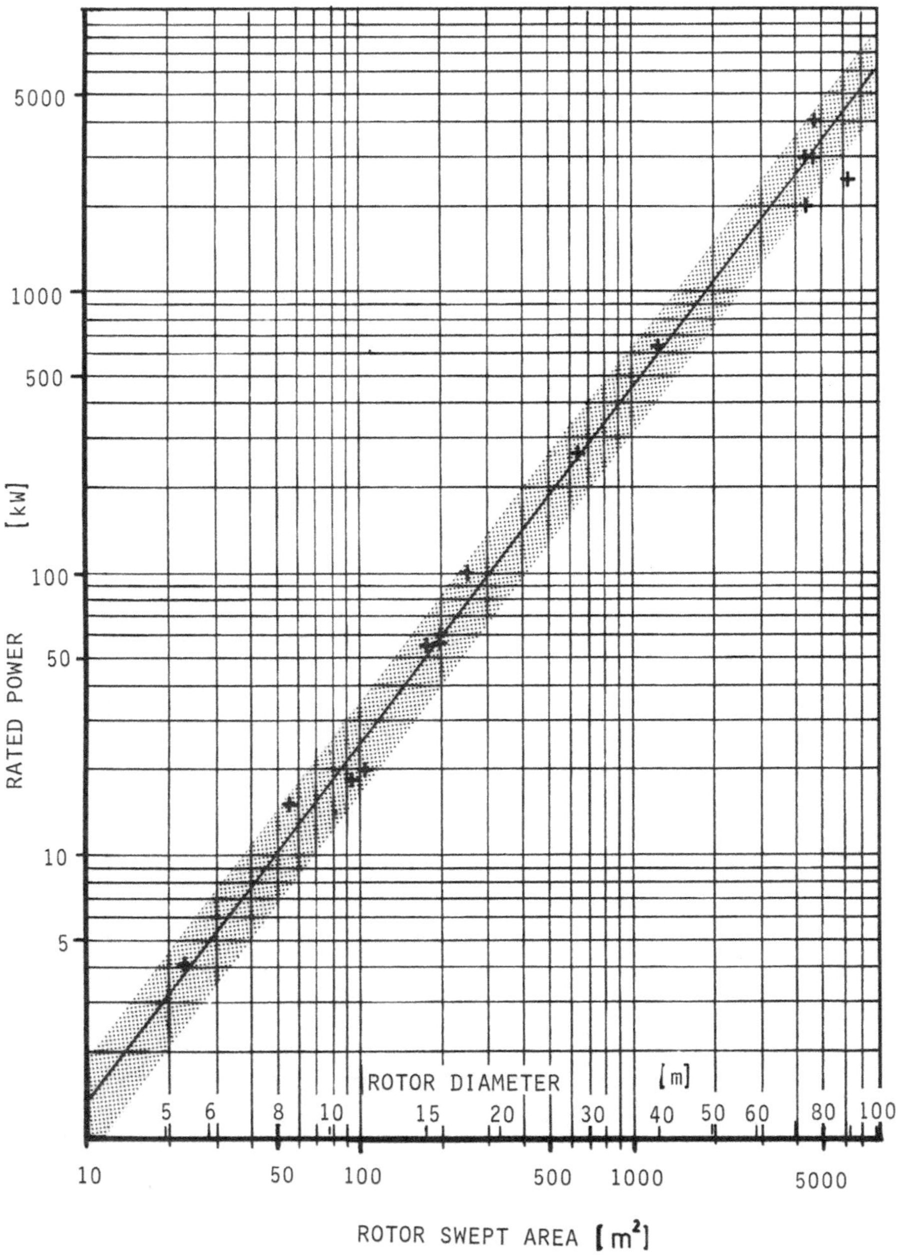

Fig. 4-1:
Typical Power Rating of Wind Turbines

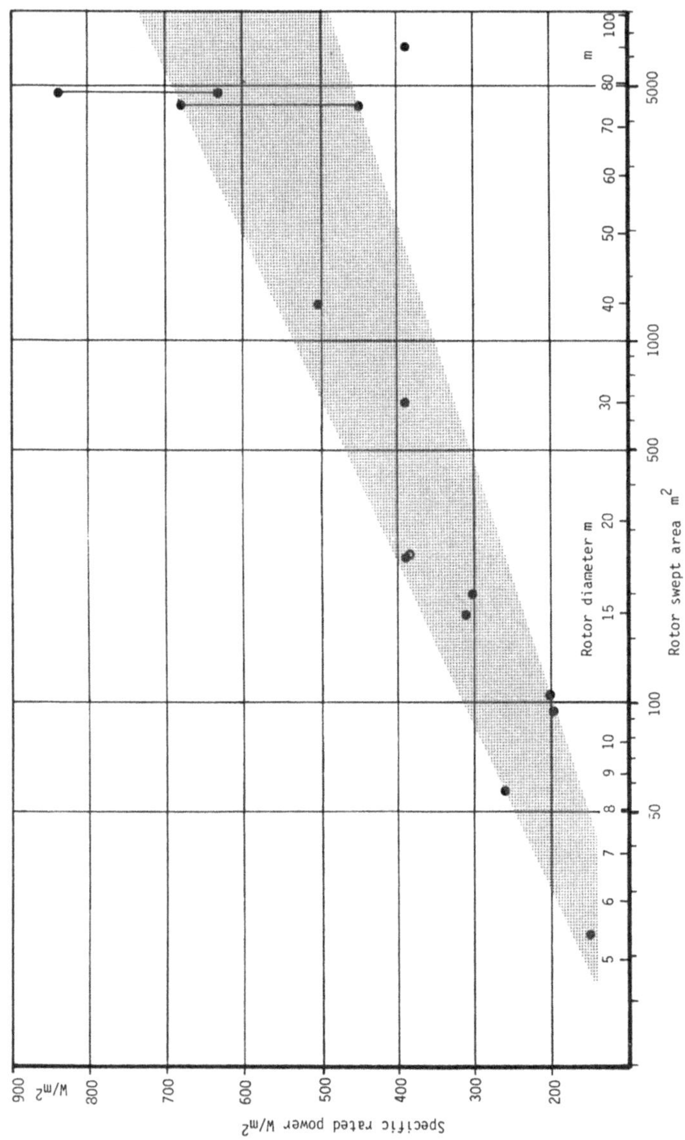

Fig. 4-2:
Typical Specific Power of Different Rotor Sizes

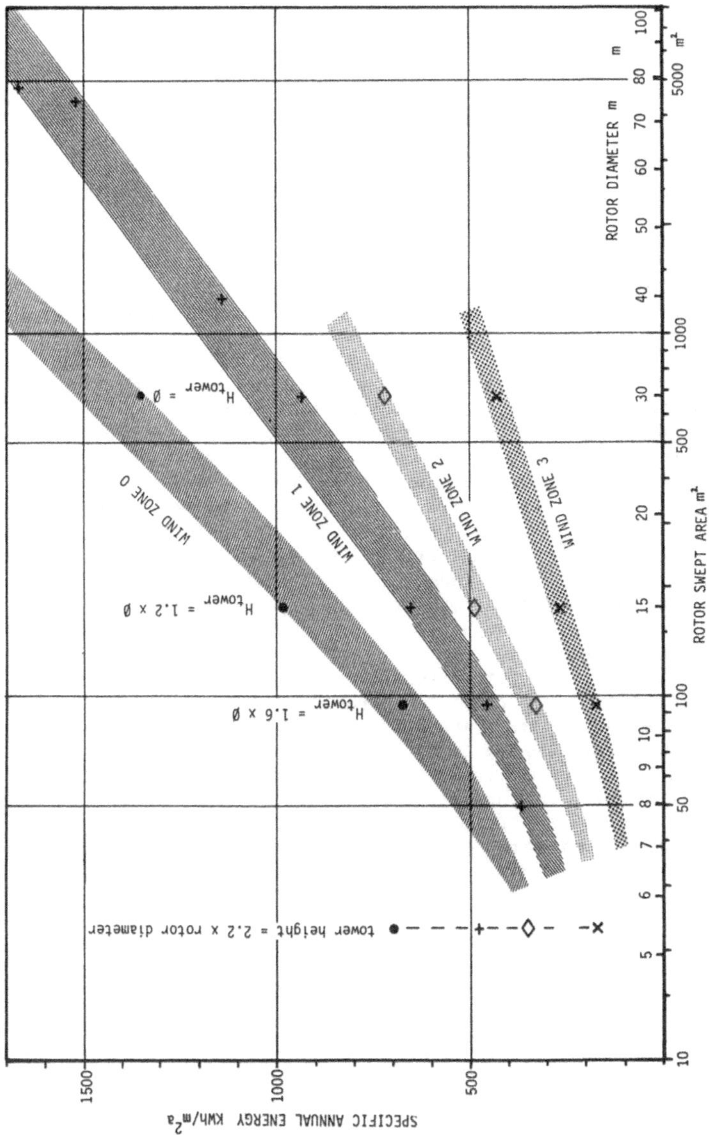

Fig. 4-3:
Specific Annual Energy of Wind Turbines at Different Sites

Although the Danish definition of the wind zones may not be applicable to areas of different climate and topography, they provide the basis for rough estimates. For example, one can expect a much higher annual energy from the Mod-2 if located in another wind zone. The energy gain in the best wind class is 4 times that of the lowest one. For economic reasons, the installation of WECs in wind zone 3 will be questionable.

Another type of presentation will be helpful for cost considerations, to be discussed in Chapter 5. In Fig. 4-4 the total annual energy produced is plotted against the rated power for those WECs which could be associated with the Danish wind classification. Although the log/log plot rules out minor differences, the good agreement over four decades exemplifies the high technical standard which has already been achieved. Dividing the annual energy by the rated power, one obtains the so-called "equivalent utilization period at maximum output capacity" which is the annual time of averaged power, corresponding to the terms used in the calculations of conventional power systems.

Searching for another systematic conformity, the next presentation in Fig. 4-5 relates the specific annual energy of the tabulated WECs to the mean annual wind speed at hub height. The lines reflect the influence of design principles and different shapes of the Weibull wind distribution curves, to mention only the most important ones. Not all of the mean annual wind speeds are based on the same wind distribution curve, hence there is a difference in the energy content underlying these average values. If we neglect the 4 kW turbine, the Danish designs show an increasing specific energy production with rotor size at the same mean wind speed. The same tendency is seen with the three Dutch machines. In principle, these results correlate with the findings of Fig. 4-3 and can be used to some degree as a measure of the design standards. Yet it must be stressed that this measure should not be taken as an evaluation of a machine, as finally the minimization of energy production costs is the goal and here the design standard is not the only influence.

At the beginning of this assessment study, the performances of three different WECs were taken and transformed to a hypothetical rotor with 100 m diameter as there was no other basis for such a characteristic. In order to check the validity of that assumption, the dashed line was included in Fig. 4-5. It matches well with the single values of Mod-2 and AEOLUS-2 MW and, bearing in mind the rising tendency with rotor diameter, one could expect a reference line shifted to higher values. Subsequently, it can be stated that the results of the potential for convertible wind energy in Table 2-3 are based on a non-optimistic reference curve.

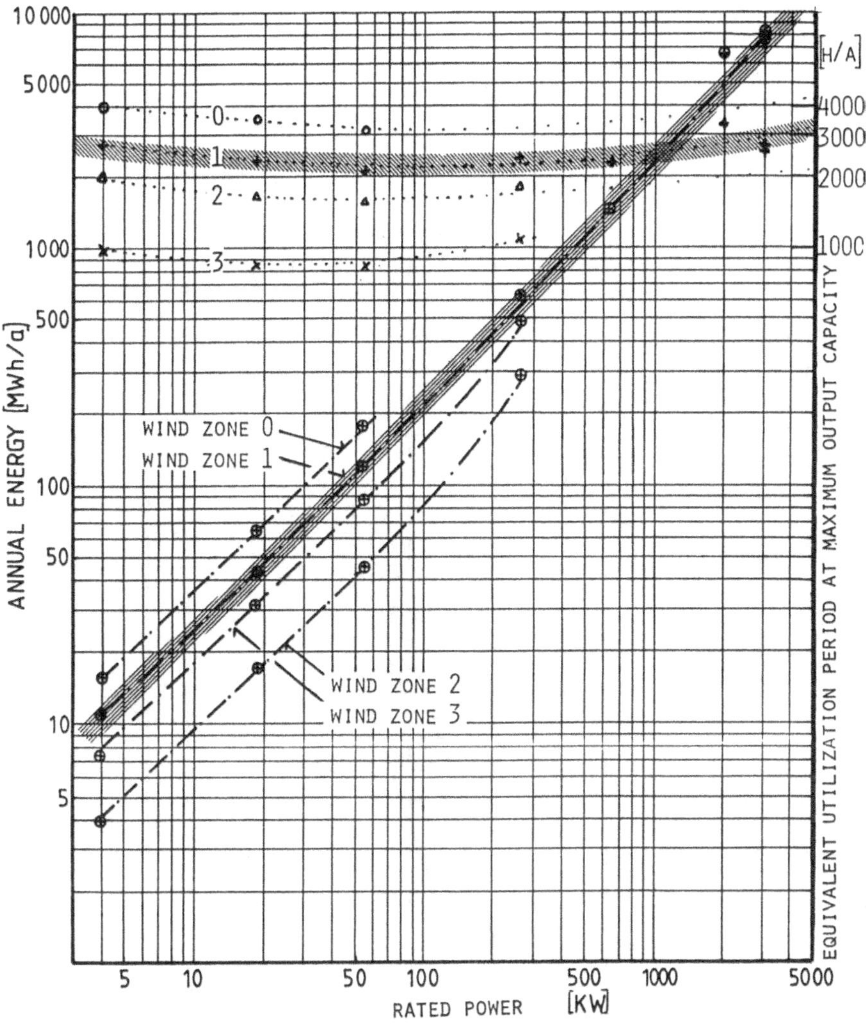

Fig. 4-4: The Annual Energy Production of Various WEC Sizes at Locations with Different Wind Potential

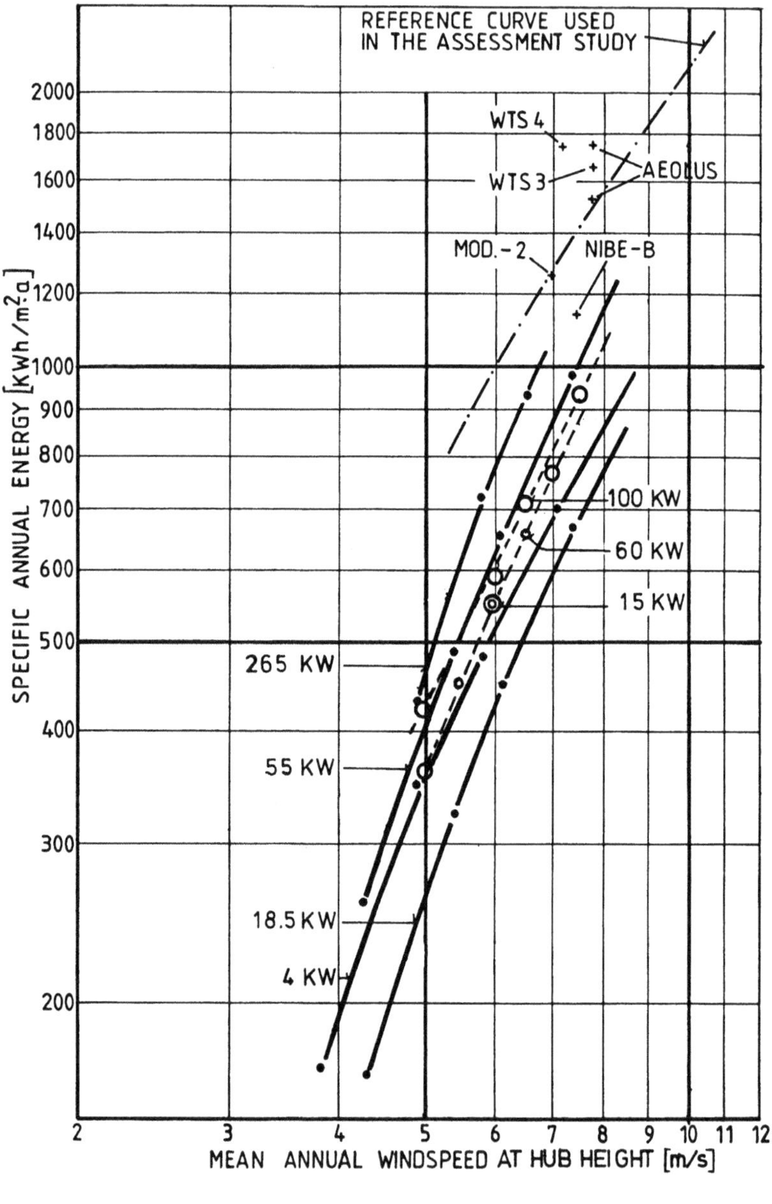

Fig. 4-5: Variation of the Specific Energy Production of Various WECs with Mean Annual Wind Speed

4.2 Integration of WECs in the Electricity Networks

4.2.1 General Aspects

The connection and integration problems of WECs differ basically with respect to the power level. Small WECs must be connected only to the low voltage grid, whereas the large turbines will feed electricity into the medium or high-voltage networks.

The low voltage grid (220/380 V) requires special attention in order to avoid or limit disturbances to other consumers. Therefore, for the connection of WECs to the grid three conditions must be satisfied:

- o The WECs must maintain the quality of the power supplied,
- o the grid must be protected against malfunctions of the WECs,
- o the WECs must be protected against malfunctions of the grid.

All utilities have published specifications giving details of the power quality required of private electricity generators. In most cases, they correspond to the values recommended by UNIPEDE.

As the "WEC" power source is rather new and some 700 units have already been installed in Denmark, a brief description of the Danish rules will be given as an example:

- o Connection of a WEC to the grid must be carried out by an authorized electrical contractor; diagrams of the equipment and a written declaration must be forwarded to the power company.
- o The equipment should be disconnected from the grid automatically in case of malfunction. It follows that the WEC can supply power to the grid only when it is being fed from other sources as well. Furthermore, a manually operated circuit breaker must be included.
- o A person must be assigned to be responsible for the operation of the WEC.
- o Moreover, an agreement on cooperation between the company and the responsible operator lays down rules for switching off the equipment.

At the present time regulations of this type exist in all European utilities although they have not yet been condensed into one special form because the number of inquiries for connecting WECs to the grid is mounting gradually.

4.2.2 Electrical Requirements for Small WECs

The ability of a network to accept the parallel connection of a private generator will depend on the network's short-circuit level and the size of the machine to be connected. In most cases, individual calculations are necessary for each installation in order to know how to establish a prospective reinforcement of the network.

The main topics of interest for the supply quality are:

- o voltage changes,
- o voltage fluctuations,
- o switch-on current,
- o fault level,
- o phase balance.

As a general rule, disturbances of the frequency are very unlikely to occur.

Voltage Changes. An export of power from the private generator to the Board's network will cause a voltage rise at the point of connection. When the generator is operating without export, the magnetizing current will depress the voltage at the point of connection. These voltage changes will occur in addition to voltage changes caused by normal fluctuations in network loading and they can be calculated using the conventional equation

$$dV = \frac{P.R + Q.X.}{V}$$

where

P = Generator Power Output (watts per phase) to the network
Q = Generator Reactive Power Output (VAr per phase) to the network)
R = Resistance of Board network
X = Reactance of Board network
V = Network phase voltage.

It should be remembered that induction generators and inverters absorb reactive power and consequently the reactive power, Q, is negative in this equation. A more complete equation taking into account the quadrature component of voltage drop might need to be used in some borderline situations but this is not normally required.

The voltage of the network to which connection is sought will vary with variations in system load. The voltage changes caused by the generator, as calculated above, must also be applied to ensure that the voltage is at no time above or below the statutory limit of the declared voltage plus or minus 6 %. Much will depend on local distribution transformer tap settings, loading and balancing of the low voltage network and loading and load cycle of the high voltage network.

The voltage at the distribution company's terminal equipment at the installation with the private generator must also be within statutory limits. Consequently, the impedance used in the calculation should include the impedance of the service cable. Where existing supply voltages are near the upper limit, the scope for voltage rises caused by private generation may be limited. In these instances, depending on the generator size, it may be necessary to install a separate distribution transformer solely to supply the premises with the private generator, or to limit the export capacity of the machine.

In Denmark, experience with the WECs in operation with limited inrush current showed that the permissible voltage variations and fluctuations are within tolerable limits in most cases if a short-circuit level at the point of connection of at least 20 times the maximum (instantaneous) power of a small WEC is provided. That is if the inrush current is less than normal current and if $\Delta V < 5$ %. Of course, such a rule can only serve as a guideline. If the inrush current is not limited, a higher short-circuit level may be necessary. One Danish utility company demands for instance a short-circuit level 80 times the maximum instantaneous output, including inrush current, in such a case.

Voltage Fluctuations (Flicker). When the input motive power available to the generator varies, there may be corresponding changes in output power. With fluctuating motive power, variations in generator output result in fluctuating voltage on the network. This is particularly relevant to fixed pitch aerogenerators operating in the range between cut-in and rated wind speed. Although voltage peaks may be within statutory limits, unrestricted fluctuations could cause annoyance to other consumers. The thick line in Fig. 4-6 between A and B corresponds to the CENELEC curve shown in the European Standardization Document EN50,006 and reproduced here. This document deals with equipment to be connected at 220/240 V and is intended as a guideline for manufacturers designing equipment for general supply to the public. The document also defines a network impedance to be used in appliance testing which will result in the voltage levels shown in Figure 6 of EN50,006 for acceptable step current levels from the device. However, these voltage levels are not generally accepted as satisfactory in the U.K. even for levels on domestic service voltages, and equipment tested to EN50,006 is only acceptable in the U.K. because the impedance of the supply

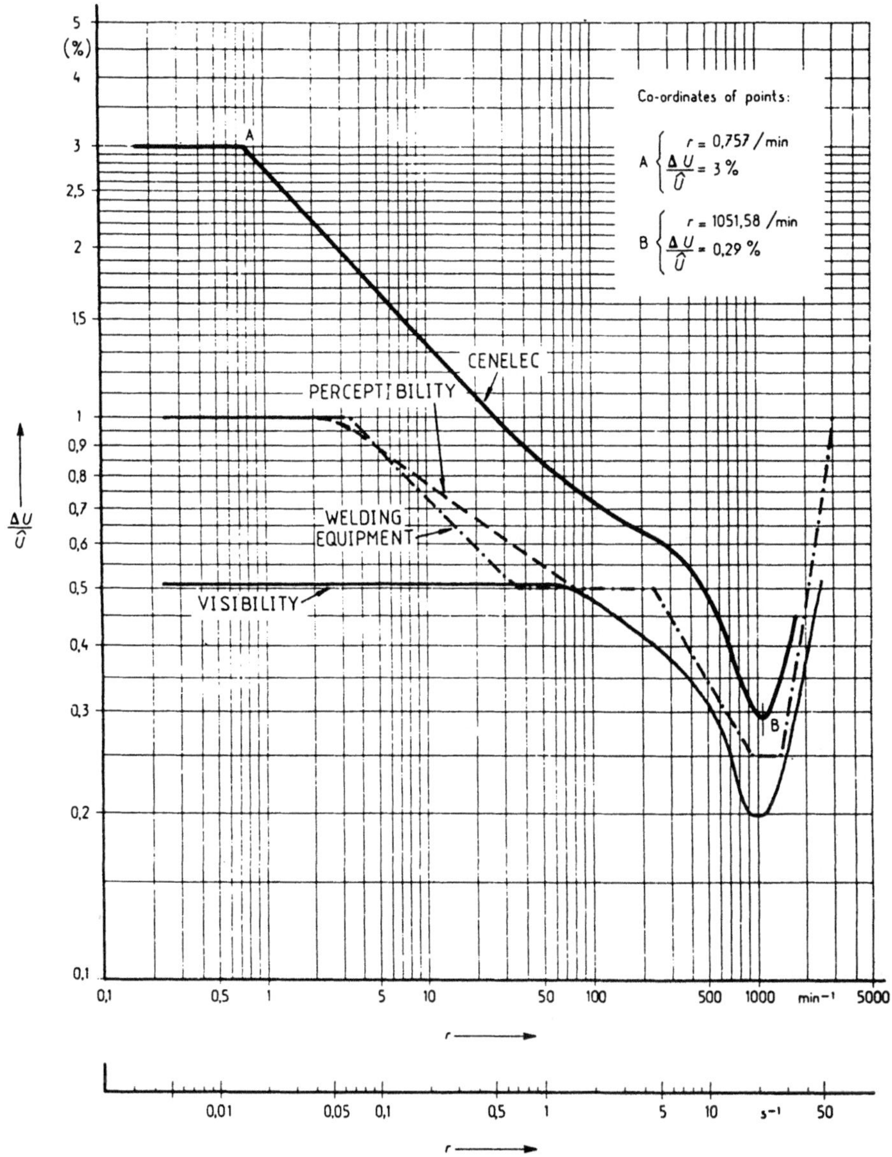

Fig. 4-6: Magnitude of maximum relative voltage changes with respect to number of voltage changes per minute (per second)

mains is generally lower, typically 0.25 + j.2 ohm compared with the value 0.4 + j.25 Ohm used in EN50,006. There seems to be some evidence that other utilities would also have some reservations about these voltage changes being applied to their systems.

The actual curve for flicker levels which can be noticed by 50 % of all people while doing other work, the perception level, is also shown on the figure and has been confirmed from recent work done as part of an activity of the "Union International Electroheat" (UIE) who are attempting to establish a standard technique for quantifying the acceptable level of disturbance from arc furnaces and other disturbing loads. The full-line section down to 2 changes/second is now well defined and is the same as the visibility limit. At lower rates the level as shown by the dashed curve is still subject to debate though recent measurements tend to confirm its validity.

It is known that the difference between the level at which interference is perceived and when it becomes annoying is relatively small. For example, a test with changes of 1.65 % at a rate of 0.1/s was claimed to be annoying by 15 % of the observers. This point lies on the CENELEC curve and is only 70 % above the level at which only 50 % of the population would notice the interference whilst working normally. In practice complaints from 15 % of the population would prove to be extremely embarrassing to the supply authority. It is also known that at the perceptibility level, although only 50 % of the population would notice interference while working, there would still be 1 or 2 % of the population who would find the level annoying. The permissible levels adopted for welding equipment also shown on the diagram were based on an assessment of acceptable levels and would have taken into account the possibility of a second welder on the same feeder.

The levels adopted for equipment connected at 240 volts do to some extent reflect the situation that there will be some attenuation in level experienced by other consumers and that there is almost complete isolation at the local area supply transformer. This, and the limited number of consumers close to one small WEC, may explain why flicker levels higher than CENELEC values as indicated in the figure are tolerated in Denmark.

The size of aerogenerator of interest to supply authorities would be connected at higher voltage levels than 240 volts. Step voltages will be transmitted to the lower voltage with virtually no attenuation and the number of consumers experiencing each disturbance will be very much greater.

The motor switching curve (see Fig. 4-7), which is probably most appropriate for assessing the starting of WTGs, is based on the assumption that there could be a significant number of other machines also contributing to the overall source of interference. The Engineering Recommendation P13/1 distinguishes between frequent and infrequent starting. For infrequent starting, that is at intervals greater than 2 hours, a voltage depression of 3 % is permitted. For frequent starting, i.e. intervals of less than 2 hours, the recommended limit is 1 % except where the build up of current is delayed when higher limits are permissible up to 3 % as shown in the curve of Fig. 4-7.

The level of permissible step voltage disturbance due to motors is under review in the U.K. mainly in respect of the application of heat pumps and it is possible that some increase in level of perhaps up to 2 % for 3 - 4 operations per hour may be found to be acceptable. However, this would still be based on the expectation that not more than two or three machines would contribute to the interference for any one consumer.

If wind power is to achieve even modest penetration into the supply system there could be many machines which are effectively close together in terms of their electrical connection. Under conditions of light gusting wind the overall frequency of cut-in disturbances could be quite high and this should be taken into account in assigning permissible step voltage limits for individual machines.

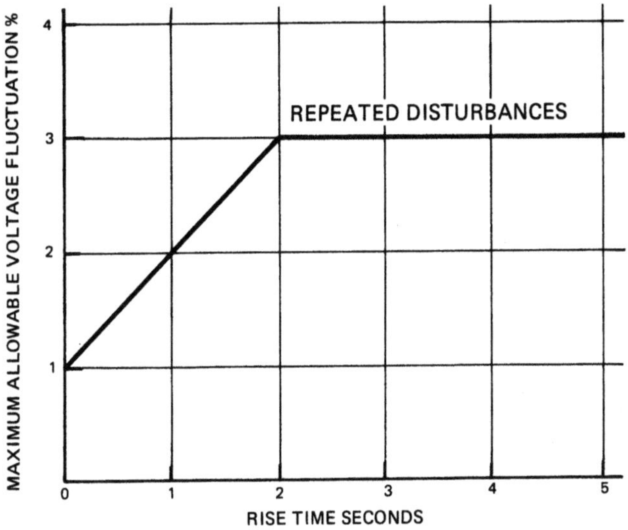

Fig. 4-7:
Repeated Disturbances: Large Motor Switching

Switch-on Current. At the instant of connection of an asynchronous generator with the utility's network there will be an inrush of magnetizing current. This may be many times the normal operating current, but will have a relatively fast decay time. This switching transient may be seen by other consumers as a flicker. The calculation of the magnitude and duration of the current is complex and differs for each machine. If the magnetising inrush current cannot be quantified at the design stage, experiment on site may be necessary.

At low wind speed, it could happen that the turbines will stop and restart several times per minute. It would perhaps be tolerable if this flicker-disturbance were limited to 3 times per hour. Minimization of the disturbance by the inrush currents can be gained by several approaches.

For example, the generator winding can be specifically designed to reduce inrush currents to minimal values. Sometimes two windings, one a higher impedance starting winding, are used. Another method is to design the mechanical coupling of the aerogenerator blades to the generator such that it is not possible for the generator to drive the blades. If during operation the wind speed becomes insufficient to drive the asynchronous generator at or above synchronous speed, the aerogenerator blades 'free wheel' and the generator runs as an induction motor on no load. A timing relay is usually included to limit the duration of this motor action to a few seconds, after which the output circuit breaker is opened. Frequent connection and reconnection of the machine to the network is thus avoided.

Furthermore, the use of AC-DC-AC inverters will also result in a reduction of these disturbances.

If no account of inrush currents is taken in the machine design it may be necessary to impose a time delay on the automatic reconnection of the generator following disconnection due to operation below synchronous speed. An alternative is to widen the difference between the speed at which the generator is connected and the speed at which it is disconnected. The low speed setting should be adjusted, as this will only result in slight motoring action with less shock to the network.

Static Balancers. The low voltage network to which the generator is to be connected may include a static balancer. During normal operation and most fault conditions the private generator will not adversely affect the balancer. However, if the Board loses supply to one of the phases to which the balancer is connected, the output voltage of the balancer remains substantially equal, and it is possible for the exported phase current from the private generator to exceed the phase rating of the balancer. Consequently, when a generator is to be connected to a network which includes a static balancer, the phase rating of the balancer should be checked to ensure that it is suitable for the maximum export capacity of the generator. Consideration should be given to fusing the neutral of the balancer. In some cases it could be necessary to change an existing balancer for one of a higher rating.

Fault Level. The limitations on voltage limits normally restrict the size of the generator operating in parallel with the network such that the private generator does not have a significant effect on the fault level of the network at the point of connection. However, if the fault level of the low voltage network is of concern, the reactance of the generator should be determined and the actual contribution to the fault level calculated in the normal manner.

4.2.3 Safety Requirements for Small WECs

As mentioned above, the governing boundary conditions for the grid connection are designed to prevent intolerable disturbances of the energy supplied by the grid and to safeguard the WEC and network. Special care must be taken if repair work is in progress on the network. The principal components of the safety subsystem are illustrated in the functional scheme of Fig. 4-9. The hardware needed for the safety demands depends on the type of generator used.

Induction (asynchronous) Generators. Normally, only mains - excited generators are allowed to be connected directly to the low-voltage network of the European distribution companies. Protection must be installed to prevent generation by self-excitation due either to the capacitance of the network or to power factor correction capacitors installed on the premises or on an adjacent consumer's installation.

Self-excited asynchronous generators must be connected directly to the Board's low voltage network.

Asynchronous generators designed to operate in either the mains - excited or self-excited modes must have circuitry arranged such that self-excitation can occur only when the generator is completely isolated from the Board's network. Suitable changeover switchgear must be installed.

Line-Commutated Inverter Connection. Synchronous, self-excited asynchronous and all other types of generator can only be connected to the Board's low voltage network by means of a line-commutated inverter. The inverter must be arranged such that commutation will cease on all phases connected. Inverters capable of operating in both the line and self-commutated modes are not acceptable.

Where aerogenerators are connected to the supply system via inverters these must comply with the limits defined in the regulations of the relevant utility company. In the United Kingdom this regulation ER-G5/3 has effectively two levels of acceptance and stage 1 is for small types of equipment and states the maximum size of converter acceptable for connection at 415 V. No consideration of system details other than the normal considerations which apply to all new connections, such as the thermal rating of the circuit or the voltage regulation are necessary.

Stage 2 is for equipment in excess of Stage 1 limits and comprises a table of maximum permitted values of harmonic currents that a consumer may feed into the system, provided the values of individual harmonic and total harmonic voltage distortion on the system are within stated limits. If filters are used to limit the output of harmonics from an inverter installation the filters must also be able to tolerate the harmonics existing on the Board's network.

Single Phase Generators. The extent to which single-phase generators can be accepted on the network is extremely restricted. A single-phase mains-excited asynchronous generator is not generally a viable machine. With very few exceptions, mains-excited asynchronous generators are three phase. Self-excited single-phase generators, both synchronous and asynchronous are available but, as with three-phase machines, can only be operated in parallel with a Board's network when connected via a line-commutated inverter.

In the United Kingdom, under the Stage 1 acceptance, single phase converters, which theoretically produce no even harmonics, are limited to 5 kVA at 240 V. The use of single phase converters which do produce both odd and even harmonics is deprecated. Larger units are not permitted because of the unbalance caused to the fundamental three-phase voltages as well as the increased penetration of unbalanced triplen harmonics to different voltage levels.

General Guidelines. General guidelines have been published on the principal requirements for the protective equipment. As the principles are the same everywhere, the following notes, derived from the British "Engineering Recommendation" G26 [4-3] and G47 [4-4], are mostly applicable.

Relays must derive all necessary supplies from the circuit to which the generator is connected. Relays relying on batteries for sensing or for operation of the output circuit-breaker are not permitted. The protection circuits must be connected such that failure of the trip circuits or the relay circuits automatically results in the opening of the output circuit-breaker and the complete shutdown of the prime mover. Visible indication of the operation of a relay is required, and all relays must be accessible from ground level.

Power factor correction capacitors increase the possibility of a normally mains-excited asynchronous generator being capable of generation by self-excitation during supply interruptions. Consequently, although obligatory protection is required to prevent continued generation, the installation of power factor correction capacitors should, whenever possible, be avoided at the generator installation or on low-voltage distributors with which a private asynchronous generator is to operate in parallel.

Every installation which includes a private generator operating in parallel with the Board's supply must include an isolating switch capable of disconnecting the whole of the consumer's installation from the Board's network. The switch must be lockable in the off position only. Locking should be by separate padlock.

The main components to be mentioned are for:
voltage protection,
voltage unbalance protection,
phase symmetry detection,
frequency protection,
overspeed protection,
earthing,
lightning.

In detail, the components must fulfill the following requirements (the figures given refer to the CEGB network):

Voltage Protection. The generator output circuit-breaker must be such that it can only close or remain in the closed position when all of the phases of the circuit to which it is connected are energized. No volt protection must be provided so that loss of supply to one or more phases causes the output circuit-breaker to open and isolate the generator from the Board's network. When normal supply is restored to all the phases, automatic reclosure of the circuit-breaker is permitted, subject to the output of the generator being satisfactory.

To prevent generation if self-excitation occurs, over- and under-voltage relays open the output circuit breaker, shut down the generator and bring the prime mover to rest if the voltage exceeds the range of declared supply voltage plus or minus 6 % for a period of 6 seconds or as specified for the local Board's network. Reclosure of the output circuit-breaker must not be possible until the voltage relays have been manually reset.

In some installations, the generator and output circuit-breaker could be remote from the Board's terminal equipment and experience voltages different to that at the intake position. When the voltage difference is significant a greater voltage range may be possible providing the voltage at the Board's terminal equipment is not permitted to exceed the statutory requirement of declared voltage plus or minus 6 %. Where such relay settings are required, the written approval of the Electricity Board must be obtained.

Voltage Unbalance Protection. To protect against continued generation after the failure of one phase of the Board's supply, relays must be installed to monitor the voltage symmetry of the output and to open the output circuit-breaker, shut down the generator and bring the prime mover to rest if the voltage unbalance exceeds 3 % for a period of 6 seconds. For this purpose voltage unbalance is defined as:

$$\text{Percentage unbalance} = \frac{\text{negative phase sequence component}}{\text{positive phase sequence component}} \times 100$$

Where longer time delays are used for voltage and frequency protection the voltage unbalance protection can adopt the same setting. Shorter time delays should not be used.

Reclosure of the output circuit breaker must not be possible until the voltage unbalance relay has been manually reset. The integrity of the incoming phases should be checked before resetting the relay. The consumer should inform the Electricity Board of any irregularity in the supply.

Phase Symmetry Detection. If only one of the three phases of the network is being disconnected, a phase symmetry relay should be activated.

Frequency Protection. In some cases, capacitor batteries installed in some part of the grid might provide enough reactive power to maintain a frequency and voltage unstable production by the induction generators, even if this part of the grid were disconnected from the mains.

To prevent such self-excited generation, over- and under-frequency relays must be installed to monitor the frequency of the output and to open the output circuit-breaker, shut down the generator and bring the prime mover to rest if the frequency exceeds the range of 48.0 Hz to 50.5 Hz for a period of 6 seconds or as specified for the voltage protection. Reclosure of the output circuit-breaker must not be possible until the frequency relay has been manually reset.

Overspeed Protection. The speed of the machine prime mover must be monitored and automatic means employed to open the output circuit-breaker, shut down the generator and bring the prime mover to rest if the speed significantly exceeds the normal range of operating speeds. The protection setting must prevent any possibility of machine runaway, and reclosure of the output circuit-breaker must not be possible until the overspeed protection has been manually reset.

In many WECs this provision is part of the control system of the turbine.

Earthing. When the Board provides an existing installation with a direct earth, the connection of a private generator does not involve any special difficulty, and fuse protection in conjunction with adequate bonding should satisfactorily protect the installation. When an existing installation is provided with PME protective multiple earthing, caution must be exercised before extending the installation to include a private generator as there may be some difficulty in achieving the necessary bonding. For example, aerogenerators may be on metallic structures remote from the premises. In these cases consideration should be given to controlling the particular plant in difficulty by a residual current-operated device and all the metalwork associated with the generator should be segregated from the rest of the installation. Where segregation is not possible, then the alternative of using residual current-operated devices (with time delays to achieve discrimination when appropriate) for the whole installation should be considered.

Small- and medium-sized asynchronous generators are normally three-wire machines. As mains-excited asynchronous generators cannot operate independently from the Board's network it is acceptable that the generator winding is not earthed. Subject to the precautions outlined above, the metalwork associated with the generator can be connected to the earth system of the installation in the same manner as conventional apparatus and consequently the consumer is not required to install an independent earth system. However, with most machines the control and protection equipment for the generator require connection to the neutral and the three phases of the Board's network. Self-excited generators may only be coupled to the network by a line-commutated inverter but may then be capable of operating in isolation from the Board's network. Consequently, when a.c. generators of this type are used it is necessary for the star-point of the generator to be earthed, and the customer should install an independent earth system. This earth system must provide approximately twice the operating current of the earth fault protection. The consumer's independent earth system can be interlinked with the earth system provided by the Board by means of an earth bond of adequate size for the consumer's installation. It should be noted that the earth connection and the wiring to the point of connection with the Board's system could carry power system fault current.

If more than one generator is to be operated in parallel at one location, the decision to couple the neutrals will be dependent on the amount of third harmonic circulating current which will result from the coupling. In such cases the use of a static balancer to provide the neutral and earth connection for the four-wire system will mean that the neutral of the four-wire system will be maintained and earthed irrespective of the number of generators connected. The generators can then be connected as three-wire machines. With self-excited machines, the method of connecting the neutral at the particular installation will determine whether the output circuit-breaker and the Board's isolating switch should be 3-pole or 4-pole.

Lightning. Structures such as those used to support aerogenerators may be prone to damage by lightning. The specific requirements differ in detail depending on the local situation. As an example, the "British Standard Code of Practice CP 326" is cited although it is currently being revised. In its present form it requires that the earth resistance of a lightning protection system does not exceed 10 ohms, but does recommend common earthing with other services and goes on to stipulate that the earth resistance should then be the lowest required by any of the regulations for such services.

However, with aerogenerators, not withstanding any connection between the lightning and electrical earths that might exist, it is desirable that a separate lightning earth electrode is installed physically and electrically as close to the prime mover as possible, and having an earth resistance at least as good as that of the electrical earth. The Electricity Board must always be consulted when consideration is being given to bonding any lightning protection system to the earth terminal of the installation.

Grid supervising relays. In some countries, equipment has been developed in accordance with the limits specified by the Power Boards. If such components are used in the installation, the Power Board will not require a proof of the specifications. As it is likely that aerogenerators from other countries will be installed, the requested operational conditions must be ascertained. The most important limits are given in Appendix 5.

In order to simplify the protection devices, a so-called "black box" has been developed and is undergoing test now. The specifications of this grid supervising relay are given below.

NESELCO's new Grid Supervising Relay 601E0700 has been developed in cooperation with DEFU (Research Association of the Danish Electricity Supply Undertakings). The relay is specially suited to minor power stations, based on asynchronous generators. These stations are characterized by only being able to supply electric power to grids already being supplied from other sources. According to a note issued by DEFU on 10 March 1978, one of the conditions for delivering electric power to a public grid is that relays supervising voltage, frequency, and symmetry, are being used. NESELCO has combined all these relays with firm factory-calibrated limits in one relay, the 601E0700.

Concurrently with the increasing installation of WECs, consideration will be given to decreasing the under-frequency limit to 48 Hz instead of 49 Hz as mentioned. The WECs will thus be able to help the rest of the generating system in sustaining production during emergency situations, instead of tripping in a crucial situation with the frequency between 48 and 49 Hz.

DATA (Relay 601 E 0700):

Voltage and frequency:	3 phases, 380/220 V, 50 Hz
Relay consumption:	2 VA
Limits:	
excess voltage:	115 %
phase symmetry:	15 %
excess frequency:	51 Hz ± 0.1
under-frequency:	49 Hz ± 0.1
time-delay:	10 sec.
Output:	Normally activated relay with reversing contact. Contact data: 250 V, 7.5 A.

The limits are firm adjusted and calibrated. Testing of these limits can be performed during operation by means of a built-in test circuit with

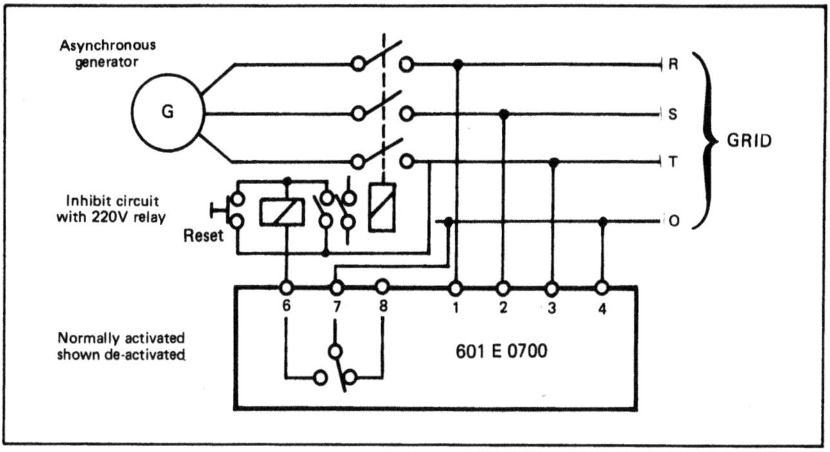

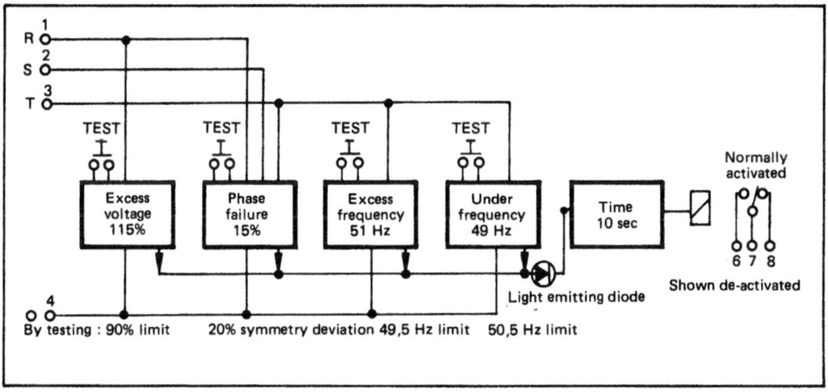

Fig. 4-8:
Grid Supervising Relay 601E 0700

special test buttons, e.g. periodically carried out by authorized personnel.

Functions are illustrated in Fig. 4-8.

Functional Areas:

Temperature:	-20 °C - 55 °C
Voltage:	-50 % - +15 %
Frequency:	42 Hz - 65 Hz

Design:

Encapsulation:	Knockproof ABS plastic (MIX box)
Dimensions:	230 x 115 x 67 mm
Weight:	900 grams
Connection:	Screw-clamps for 0.5 mm^2 to 2.5 mm^2

4.2.4 Safety Requirements for Large WECs

The limit for the connection of small WECs to the low voltage network lies in the order of 100 kW rated power. Turbines with higher power rating must be connected to the medium-voltage (MV) or high-voltage (HV) networks.

The stability and power transmission capacity of the MV and HV grids compared with the "medium WEC" and "large WEC" power sources eliminate many possible concerns of the Electricity Boards, especially as these units will be operated by the supply authorities themselves. Furthermore, it can be assumed that the supply and connection of the electric subsystem for the WEC will be carried out by a company experienced in the installation of high-voltage equipment. This implies that already at the design phase of the electric subsystem a close link between the Electricity Board and the contractor for the large WEC exists - and in most cases the Board is the contractor.

On the other hand, the development of large WECs has not yet reached the stage of series production, therefore electrical requirements do not exist to cover all locations but will be established in an individual way by the local Board concerned. As only a few large WECs have been installed up to now, an example is given of the requirements for the Swedish large wind turbines, as specified in the "Technical specification" of the Request for Proposal by the Swedish Board of Energy Production Research (NE) in 1979.

Electrical network:

- o nominal frequency — 50 Hz
- o frequency variation
  - normal — ± 0.1 Hz
  - at fault conditions,
    continously — ± 1 Hz
    rate of variation — 0.5 Hz/s
  - at extreme fault
    limited duration — ± 2.5 Hz
    rate of variation — 1 Hz/s

Note:
Lower and higher frequencies may occur for considerable intervals. The electrical network is then considered as being out of function.

- o Voltage variation
  - normal operation — ± 5 %
  - at fault condition — -15 %
    (long duration) — +10 %

Note:
Lower voltage and/or loss of voltage may occur.

- o Short-circuit power level at the high-voltage connecting point
  Gotland: min. 40 MVA, max. 250 MVA
  connecting line voltage 30 kV

- o Insulation level

  According to SEN 210510.

Electrical Requirements

- o Normal operation

  The reactive power consumption from the network must be kept low and under control. If an induction generator is chosen, the reactive power need shall primarily be satisfied by internal capacitors. The maximum voltage variations allowed at the network connection due to the WTS operation are:

  - transients at switching the WTS on and off the network ± 5 %
  - slow variations, or continuously ± 0.5 %

Note:

The requirements apply to the high-voltage side of the transformer.

- o Fault conditions

  The protective devices of the WTS shall disconnect the WTS from the network within 0.2 s at internal electrical faults.

  The WTS shall be disconnected from the electrical network within 0.2 - 2 s at a short-circuit and 5 s at an earth fault in the nearest radial section of the network.

  The WTS shall be kept connected to the electrical network at fault conditions at other parts of the network than the nearest radial section, with the following operational conditions, except in CASE 1 which shall be adhered to when possible.

Case	Frequency (Hz)	Rate of variation (Hz/s)	Voltage (%)	Duration
1	49 - 51		0 - 85	1.0 s
2	49 - 51	0.5	90 - 105	Cont.
3	47.5	1	95 - 105	30 min
4	49.7 - 50.3		85 - 90	1 h
5	49.7 - 50.3		105 - 110	1 h

  At operational conditions outside what is defined in the above table, the WTS shall be disconnected from the network.

- o Protective devices

  Electrical and mechanical devices shall protect the WTS and disconnect the WTS from the network in case of abnormal conditions in the WTS or in the network, as defined above.

In view of the controlled operation of large WECs by the supply companies, the connection of synchronous generators can be justified when adequate protection facilities are installed. Conventional equipment in standardized versions is now available for all requirements of the electrical subsystem.

As the components for the electrical subsystem were developed for other applications, it is obvious that they might not represent the optimum solution for large WECs with respect to economy. Therefore, studies and development work have been started with the object of producing equipment specifically designed to ease the operation of large WECs and improve energy production.

4.2.5 Problems of Island Operation of WEC and Diesel Systems

Most WECs have been designed to operate while connected to a strong network which stabilizes the frequency and voltage. Small networks on islands or in remote areas are supplied by diesel generators with very high generating costs. The satisfactory, reliable operation of wind turbines combined with diesel plants would be very attractive to many small networks located in wind-rich areas.

Two prototype studies have been initiated:

- o the installation of 5 Aeroman Units (20 kW each) on the Isle of Kythnos within the German-Greek coöperation programme.
- o the installation of a 63 kW wind turbine on the Aran Islands of Ireland under a contract of the Department of Industry and Energy.

As the programmes of experimental evaluation are not yet concluded, a brief description of the Irish project will be given to illustrate the problems involved.

Inis Oirr is one the Aran Islands, lying some 50 km south west of Galway city, in the Atlantic. It has a population of approximately 300 who are served by a local 44 kW diesel-electric generator. A 26 kW standby generator is also installed, but is not capable of parallel operation with the 44 kW set. Power is transmitted at 10 kV single-phase 50 Hz and distributed at 220 V. Annual consumption is about 80,000 kWh. The exact wind climate is not known, but an annual average wind speed at 10 m of 7 m/s seems likely.

Following feasibility studies and detailed designs, the Department of Industry and Energy installed a 63 kW synchronous wind generator on the island in November 1981 at an elevation of 60 m near the highest point of the island. A 12 kW diesel generator was also installed in the existing generator building, 800 m away, and it is capable of parallel operation with the wind generator as a "peak-lopper". This "package" is not designed for parallel operation with the existing diesel and no storage is provided. The wind generator and building are connected by a 10 kV three-phase conductor for ease of synchronization and control and to reduce transmission losses, and also by a multi-core signal cable.

The wind generator, which has fixed-pitch blades equipped with air brakes and spoilers, disposes of excess output into a stepped resistive air-cooled load located at the generator building in order to maintain voltage and frequency. If the frequency falls, the resistive load is shed and if frequency then cannot be maintained, the 12 kW diesel is automatically started, synchronized and switched in for at least 15 minutes. A minimum load of 3 kW is applied to the 12 kW diesel at all times, if necessary by the dump load, partly to reduce damage caused by idling and also to permit fast pick-up of frequency if the net load suddenly increases. Subsequent rapid frequency control is provided by the diesel's governor.

If the wind generator "package" cannot maintain frequency, the existing diesel is automatically started and run up, the "package" disconnected and the 44 kW diesel switched in. The change-over period during which all power is lost on the island is 3/1000 second. The larger diesel then runs for at least 30 minutes and any power generated by the package is dumped. All controls on the system are conventional electro-mechanical or solid state devices and indeed all the equipment installed is commercially available.

Brief specification of the wind generator package:

Wind turbine:

Manufacturer:	Windmatic and Clarke Chapman
Hub height:	15.5 m

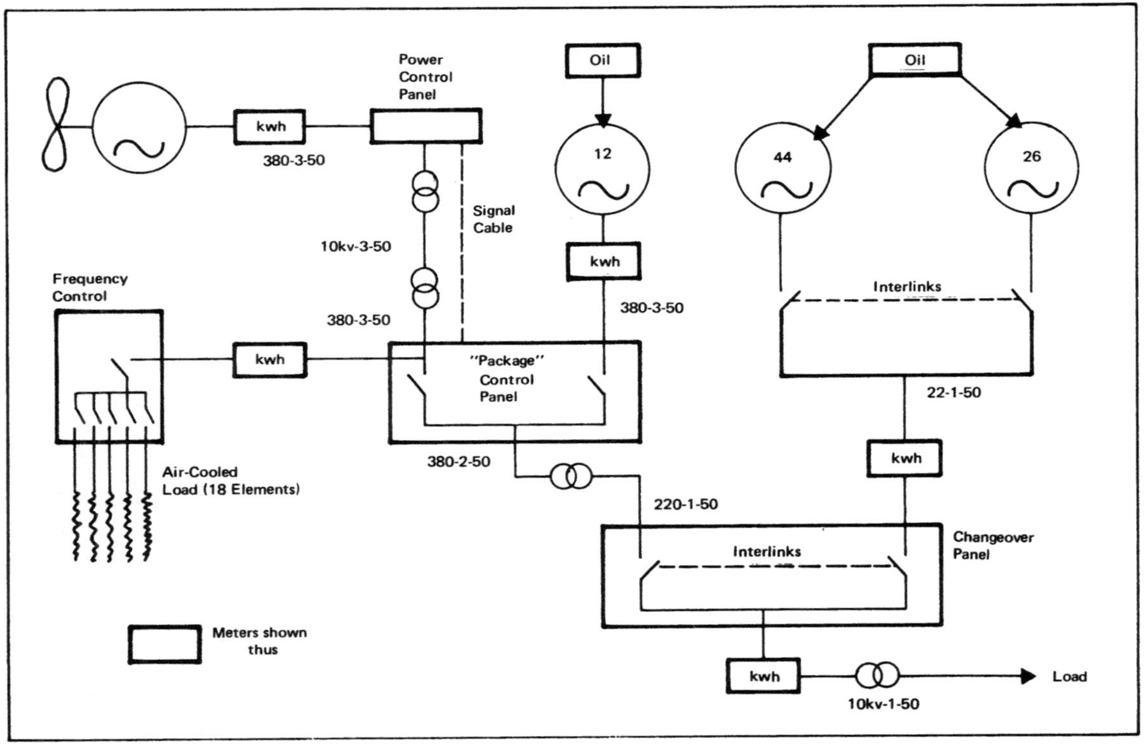

Fig. 4-9:
Inis Oirr Project - Schematic Diagram

Blades:	3, laminated wood spar with GRF aerofoil, fixed pitch, upwind.
Diameter:	14 m
Chord:	950 - 1430 mm
Yaw control:	by two tail rotors
Alternator:	380-3-50, synchronous, 63 kW
Tower:	self-supporting trussed steel
Alternator (diesel):	380-3-50, synchronous, 12 kW

Monitoring - the following will be recorded on magnetic tape: wind speed, direction, temperature, humidity at 10 m, wind speed at hub height; , wind generator output. Outputs from the 44, 26, and 12 kW diesels will be logged daily, as will oil consumption.

A schematic circuit diagram is shown in Fig. 4-9.

4.3 National Programmes

The renaissance of wind energy started in 1973, following the first oil crisis. Some countries had a clear recollection of wind energy utilization because they stopped their turbines only some 10 to 20 years ago, some countries still supported small scientific groups of research facilities, and there were even several companies producing small aerogenerators for special applications.

The real interest emerged up when the USA initiated a governmental programme to examine the possibilities of using wind energy as a major energy source and encouraged the endeavour in some European countries as well to revive their national interest in wind energy.

It is obvious that many different influences, resulting from tradition, the national energy reserves and energy programmes etc. had their different impacts on the initiation and extent of national wind energy programmes.

All the countries of Europe-10 now have official interests in wind energy.

Belgium

The Belgian National Energy Programme includes a project for the design and optimization of a 50 m diameter HAWT (horizontal axis wind turbine). This project was planned to start in January 1983.

Current programmes consist of action by the Institute of Scientific Research in Industry and Agriculture (IWONL-IRSIA), who finance the development and research of a 22 m diameter electricity generating HAWT, a 10 m diameter heat-generating VAWT (vertical axis wind turbine) and a 5 m diameter HAWT; and by the General Agency for Cooperation and Development (ABOS-ATCD), who sponsor the research on a high-solidity straight-bladed 6 m diameter VAWT for developing countries.

Other industrial development projects are sponsored by the Ministry of Economic Affairs and by the Ministry of Public Works.

Denmark

The Danish programme started in January 1977 with three projects:

- o refurbishment of the Gedser turbine with respect to measurements and analysis of structural problems;
- o planning, engineering and construction of two large, electricity-producing wind energy conversion systems;
- o investigation of location possibilities.

The main support comes from the Ministry of Energy and the electric utilities (in the case of large WECs, the financial share is 80:20).

The design and erection of the two Nibe turbines has been the main activity so far, but other activities such as tests, measurements, site investigations, theoretical studies, manufacturing of wooden blades etc. have also played a major role.

The two Nibe turbines have the following main specifications:

Rotor diameter: 40 m
Hub height: 45 m
Gearbox: 3-stage, 1:45
Generator: Induction
Electrical rating: 630 kW, 6 kV

Both turbines have 3-bladed upwind rotors. The blades of turbine A are supported by stays. Turbine B has self-supporting blades. All blades have fibreglass shells attached to steel/fibreglass spars. Turbine A has fixed pitch blades. The blades of turbine B are pitch-controlled.

Turbine A was put into operation late 1979, and in November 1980 it was shut down for a major inspection. This revealed the need for new stays of an improved design. After other repairs the operation was resumed in March 1982.

Turbine B was put into operation during the autumn of 1980, and it was taken out of service in June 1981. Because of a fatigue crack detected on one of the blades, all three blades were taken to the workshop for reinforcement of the load-carrying steel spars. The operation was resumed in May 1982.

As of October 1982, the operational hours and accumulated energy production of the A turbine were 2,280 hours and 465 MWh. For the B turbine the corresponding data were 2,210 hours and 470 MWh. The estimated annual energy output is about 1.5 GWh per turbine at the site, where the average wind speed at hub height is 7.3 m/s.

Based on the experience gained from the Nibe wind turbines, the next step has been taken which is to develop a prototype machine with the following main specifications:

Rotor diameter: 60 m
Hub height: 60 m
Gearbox: Planetary, 1:67
Generator: Induction
Electrical rating: 1.5 - 2.0 MW.

The rotor is a 3-bladed upwind rotor. The blades are made of a fibreglass composite throughout. Power regulation is by full span pitch control.

Funding is presently available for detailed design and for the manufacturing of one test blade (30 m).

A major impetus to private wind energy utilization came from the establishment of a test centre for small WECs at the Risö National Laboratory.

The investigation of siting problems led to the development of a general method to calculate the local wind energy based on the tables in the "Wind Atlas for Denmark".

In addition to the activities sponsored by the above-mentioned wind power programme, other Danish achievements might be mentioned.

The Tvind turbine is well known. It has a 54 m diameter downwind rotor and a variable speed 2 MW generator. For various reasons, maximum output power is presently 900 kW. The accumulated running time is 7-8000 hours.

A private company has developed a 265 kW wind turbine which is presently undergoing commissioning tests. The rotor diameter is 29.3 m, and it is a downwind machine with 3 fibreglass blades.

Part of the national programme is to stimulate private initiatives by providing assistance in the early marketing phases. In this way, the private wind turbine industry has sold more than 700 small wind turbines (smaller than 100 kW) to Danish customers, who are encouraged to buy by a government subsidy of 20 - 30 % of the purchase price if the units have been tested successfully at the Risö test centre.

France

The French activity in wind energy has a long history. Several companies produce small WECs for special applications in remote areas, such as battery chargers, etc.

The national programme directed by Agence française pour la maîtrise de l'energie has three goals:

- o the development of small WECs,
- o the development of a twin rotor with 40 m diameter,
- o the installation of a WEC test centre.

Wind energy is utilized on the islands and in the French territories outside the mainland where power generation costs are very high.

In the twin rotor concept, it is intended to mount two separate rotors on a cross-bar at the top of one tower.

West Germany

The German wind energy programme comprises 3 elements:

- o large wind energy converters,
- o small wind energy converters,
- o basic research and development.

Within the large WEC programme, four different rotor systems have been developed:

- o Voith WEC 520, 265 kW, rotor diameter 52 m, located at Stötten near Stuttgart, first grid connection in March 1982;
- o MBB Monopteros, 370 kW, single-bladed rotor of 48 m diameter, located at Bremerhaven, first rotation in December 1982;
- o MAN Growian, 3 MW, rotor diameter 100.4 m, located at Brunsbüttel near Hamburg, assembled in December 1982;
- o Thermal Upwind Power Plant, 50 kW, chimney height 200 m, solar heated intake area 250 m diameter, first tests in July 1982; engineering model for large power plants.

The programme for small WECs is concentrated mainly on:

- o the establishment and use of the test field on the isle of Pellworm,
- o the development of a vertical axis rotor within the scope of the Argentine-German cooperation programme (20 kW, 12 m diam.), first test run in May 1981;
- o operational experience of small WEC, connected to small diesel power stations; joint Greek-German project on the Greek isle of Kythnos, using 5 WECs of 20 kW power each;
- o the development of a 3-bladed rotor (25 m diam., 100 kW) within a German-Brazilian cooperation programme.

The basic research and development programme has covered

- o advanced concepts for large rotor blades,
- o control strategies,
- o wind mapping,
- o siting problems,
- o new conversion concepts,
- o theoretical treatment of static and dynamic behaviour of WECs,
- o installation of a special wind tunnel capable of simulating a gusty environment.

In addition, the German company MBB/VFW-ERNO is cooperating in the Swedish programme and has participated in the development, production and testing of the 2 MW-WEK AEOLUS (75 m diam.) located in Gotland (first test run February 1983).

Greece

There is a publicly expressed political will for the utilization of alternative energy sources and especially wind energy, but at the moment there are no definitive energy goals. The Research Administration (YEET) has, in principle, accepted the sponsorship of sixwind energy research programmes covering various aspects of wind energy (both horizontal and Darrieus type wind machines); however, these projects have not yet been funded and started.

As to the practical of utilization of wind energy: the Public Electricity Company, in collaboration with the German firm MAN under the sponsorship of the German Ministry of Technology has installed five horizontal axis WECs in the Greek island of Kythnos (in an energy park) which have the following characteristics:

- rated output: 20 kW
- rated wind speed: 14 m/s
- type of generator blade pitch control: asynchronous
- diameter: 11 m

These units were in parallel operation with the local electricity grid in May 1982 and the experience gained has given an impetus to the efforts of the public electricity corporation for the installation of larger wind generators.

The Greek Electricity Generating Board has recently started a programme to record wind speeds in various sites specially chosen for wind power applications.

Activities in research and development have been pursued since 1978 at several universities and they provided the basis for the first Greek conference on wind energy, held in Athens, October 1980, with approximately 30 Greek scientific papers.

Ireland

In 1980 the Minister of Energy launched a Windpower Demonstration Programme which included the erection of 12 different WECs serving quite different purposes. The main thrust of the programme was to be the erection of proven commercially available WECs rather than on R&D leading to the production of new machines. Certain specific areas of research were seen to be important.

The machines which have been selected are:

Location	Use	Power kW	Converter	Hub Height m	Diameter m	Blades/ Pitch
1. Sligo	Heating	10	AC Variable voltage	9.3	6.0	5 and 3 contra rotating
2. Leitrim	Heating	10	DC Variable voltage	11.2	7.6	2/variable
3. Kinsealy	Heating	10	AC Variable voltage	16.0	7.0	16/fixed
4. Fermoy	Heating	10	Water Brake	19.0	7.5	3/variable
5. Ballinrobe	Drainage	1	Mechanical	7.0	2.6	4/fixed
6. Malin	Radio	0.8	DC	8.0	3.0	2/fixed
7. Aran	Island Supply	63	AC fixed frequency	15.5	14.0	3/fixed
8. Bellacorick	Utility grid	120	AC fixed frequency	35.0	18.0	2/variable
9. Miltwon	Connection	50	AC fixed frequency	26.3	17.6x 11.4	2 VAWT fixed
10. Poulaphoca	Connection	55	AC fixed frequency	16.5	14.0	3/fixed
11. Portlaoise	"Private" grid	10	DC+Inverter	11.2	7.6	2/variable
12. Ballyferriter	Connection	30	AC fixed frequency	15.0	10.0	3/fixed

Installations 8-11 inclusive were installed by the Electricity Supply Board (ESB), the other 8 machines by the Department of Industry and Energy.

A detailed analysis of the programme's objectives is as follows:

- o To establish wind speed, wind direction, and gusting characteristics for each WTG site.
- o To establish a power versus wind speed curve for each WTG.
- o To establish the annual energy production for each particular wind regime as an aid to investigating the economics of wind power.
- o To investigate the response of the different WTG control systems to the varying wind speed, direction, and gusting conditions.
- o To establish each machine's overall efficiency by investigating the aerodynamic, mechanical and electrical losses.

- o To investigate the stresses/strains on the blades, rotor and power structures both in normal and gusting conditions. This information will help in the overall fatigue analysis of each machine.
- o To investigate the electrical problems associated with switching in/out and running of each machine of the network, including:
  - Fluctuations in frequency, voltage and power,
  - The size of the inrush current and the reactive power required,
  - General stability.
- o To establish the best safety requirements for a WTG.

Number 7, the island project, is technically the most complex. The existing diesel electric supply on the island of 44 kW capacity is operated while interlinked to a 63 kW wind generator and a 12 kW diesel generator for peak demands.

R&D activities have been promoted on

- o the development of small wind turbines,
- o studies of wind power penetration problems,
- o wind resources.

Italy

The main programmes of research and demonstration in the wind energy field are supported and developed by the National Research Council (CNR), by the National Committee for the Research and Development of Nuclear and Alternative Energies (ENEA), and by the Italian National Electric Energy Agency (ENEL).

Through its "Energy Plan 2" (Progetto Finalizzato Energetica 2), CNR is conducting basic research, especially in the field of anemology and wind-turbine generator components.

CNR's activities cover the following subjects:

- o Large wind-turbine generators
  - Improvement of aerodynamic calculation systems
  - Blade manufacturing technology
  - Electric power transmission and choice of electrical machinery
- o Analysis of nationally-available wind resources
  - Method of calculating available energy
  - Production of wind maps
  - Study of a sample area
- o Small and medium-sized wind-turbine generators
  - Innovative systems
  - Technology regarding textile materials

ENEA, within the framework of activities aimed at developing alternative energy sources, is working to encourage developments in industry and to promote the application of wind energy.

ENEA's programmes are subdivided into the following headings:

- o Choice and preparation of sites
  - Pinpointing of sites in regions with good wind resources
  - Map of sites for medium and large wind-turbine generators
  - Participation in the preparation of a wind map of Italy
- o Station for testing and developing wind-turbine machines of a power $\leq$ 100 kW
  - Construction of a test station
  - Testing of wind-turbine generators harnessed to specific consumer installations
  - Development of multi-blade wind-turbine pumps
- o Medium-sized wind-turbine generators
  - Pre-design analysis of 200 kW horizontal and vertical-axis wind-turbine generators

- Commissioning of a 200-300 kW prototype for isolated systems
- o Large wind-turbine generators (in co-operation with ENEL)
  - Preliminary design for a 2-5 MW machine (Gamma Project)
  - Commissioning of a 2-5 MW prototype
  - Development of component manufacturing technologies

ENEL has, for some years now, through the VELE Project, been engaged in development and demonstration work on wind-turbine generators, with the active participation of the leading aeronautical and electromechanical companies operating in the wind energy sector.

In particular, the work carried out by ENEL can be described as follows:

- o Small and medium-sized wind-turbine generators (VELE Project)
  - Wind energy plant siting
  - Design and use of wind-turbine generators and their components
  - Design, construction, and experimental operation of a 500-kW wind power plant
  - Installation and experimental operation of wind-turbine generators for demonstration purposes in various Italian regions
- o Large wind-turbine generators (in co-operation with ENEA)
  - Preliminary design for a 2-5 MW machine (Gamma Project)
  - Commissioning of a 2-5 MW prototype

For the implementation of the above programme, overall government expenditure, as estimated by the Ministry for the Co-ordination of Scientific and Technological Research in the 1981 "National Energy Research Plan" (PNRE), would be Lit. 35,000 million (at 1981 prices) for the period 1981-1985.

The Netherlands

In the follow-up programme of the Dutch national research programme on wind energy (NOW - July 1981) it is stated:

"The objectives of the programme are to provide the development strategy towards the application of wind energy with at least a total capacity in the year 2000 of:

- 2000 MW installed in wind power stations (1,000-1,500 units),
- 450 MW installed in small-scale decentralized applications (about 15,000 units)."

This aspect raises expectations for the future.

Owing to rising energy prices and the high investment premium for using energy-saving installations, about ten Dutch manufacturers have started to produce wind turbine installations during the last two years. Mainly small-scale wind turbines (d = 8-16 m) are produced, not only for the Dutch market but also for export. These wind turbines are generally used for generating electricity, the production of heat for greenhouses, or mechanical energy in water-pumping stations.

The Government has started several demonstration projects to stimulate the introduction of wind energy in the Netherlands. So far, mostly Dutch wind turbines have been used in these projects.

Furthermore, a test station has been set up by the authorities to carry out the necessary inspections and performance measurements.

Technical and safety requirements for wind turbines are being formulated and will be given official status in the near future.

Up to now, only one Dutch company has acquired experience in the building of a large wind turbine (the 25 m HAT experimental turbine in Petten). So far, industry has adopted a waiting attitude, but it is expected that this will change in the near future: the SEP (Cooperating Electricity Generating Companies) intends to install a wind park of 10 MW within the next three years.

In this context, the decision was taken for the realization of a "wind farm" containing sufficient WECs to obtain experience with cluster effects. The WECs should be medium-sized units with a capacity of 0.3 to 0.5 MW each. The Government and the SEP consider this project as a vital first link in the development of wind energy on a large scale in the Netherlands. The demonstration project will be financed by the Government and the SEP on a fifty-fifty basis. Supervision of the realization and operation of the demonstration project is in the hands of the SEP.

United Kingdom

The UK is well placed to make use of wind generated electricity. It has a wind speed regime higher than in many other countries of the Community; large off-shore areas where arrays of machines could be sited; and, a strong national grid which is able to accept fairly high levels of non-firm power. The latter is important in the long term, although initial marked penetration may be in isolated communities with high costs of supply.

The main thrust of the research funded by the Department of Energy is aimed at the design and demonstration of MW size machines at high mean wind speed sites. To this end, three horizontal axis machines (2 x 250 kV and 3 MW) are to be erected on the mainland island of Orkney, at a hill-top site with a mean wind speed of 10 - 12 m/s. This island was recently connected to the national grid, but previously relied on a local diesel-powered station. It represents therefore that segment of the market most likely to be economically attractive in the short term.

In addition to the horizontal-axis designs, a vertical-axis machine is being developed and a decision on the building of a prototype will be taken soon. Other Department of Energy research includes estimating the resource (involving wind profile measurements and wake interaction effects), studies into the problems of offshore siting and some basic research with small machines.

The generating boards are participating in the work in their role as potential purchasers of large machines. The North of Scotland Hydro-Electric Board (NSHEB) are involved in the Orkney machines described above and will operate them. The Central Electricity Generating Board are also buying a machine (200 kW) to be sited at Carmarthen Bay and have plans for a MW generator on the east coast of England. This second scheme may lead to a cluster of ten machines.

Conclusions

The survey of national programmes reflects the increasing interest in wind energy utilization. Pace and goals differ, but in most cases national solutions are being preferred. By an enhanced exchange of experience via close cooperation, the development of viable technical solutions could be accelerated, and this would in turn enable a more rapid introduction of these new industrial products.

4.4 Involvement of Industry and Electricity Supply Boards in Wind Energy

The volume of the national programmes and the degree of involvement of industry indicate how close we might be to the introduction of WECs into the market.

Taking into account the WECs for the production of electricity and excluding non-marketable research turbines, Table 4-2 reflects the present situation in the industry:

	number of companies with line-production of WECs; in brackets: companies which have already produced a prototype		
Country	1 - 99 kW	100 - 1000 kW	> 1 MW
Belgium/Lux.	7	-	-
Denmark	22	1 (2)	-
France	8	1	-
W. Germany	7	(3)	(2)
Greece	1	-	-
Ireland	-	-	-
Italy	5 (3)	(1)	-
Netherlands	10	1 (1)	-
United Kingdom	5	(4)	-

Table 4-2:
Survey of Companies Already Involved in the Production of WECs

These figures cannot be regarded as very precise or permanent since there is some fluctuation in the companies involved in wind energy. Furthermore, there are several established companies producing small WECs for battery chargers, pumping, etc. having a power rating as low as 100 W. In this assessment study, little attention has been given to these small units, since contributions below 1 kW power cannot have much impact on the overall energy study. The situation and state of the various WECs in production and undergoing tests will be presented and discussed in detail in the Final Report of Action 4.

Despite the above reservations, Table 4-2 brings out some general features. Denmark is by far the most active country in the field of small WECs, followed by The Netherlands. But the other countries, apart from Ireland and Greece, have also developed some production capacity.

West Germany concentrates on medium and large turbines, this is also the tendency of a few other countries which follow in a step-by-step manner.

Short surveys of the situation in individual countries are given below.

Belgium/Luxembourg

There are about 3 companies producing small WECs below 15 kW rated power, and 4 companies producing small WECs between 15 and 100 kW rated power.

Several units of one type were installed as far back as August 1982 in California.

In addition, two more companies are working on the development of small WECs. With respect to the components, almost all parts for small WECs can be produced in the country.

Denmark

Danish industry is notably involved in the production of WECs. The main impetus comes from small and medium-sized companies which are aware of the market opened up by governmental subsidies.

About 40 different types are being produced by approximately 22 companies. One company is manufacturing a turbine of 265 kW rated power. A utility company installed one of these in southern Jutland.

Danish industry supplies most of the components needed for the small and medium WECs, and has started to export complete WECs as well as rotor blades in a non-expansive glass-fibre technique.

The utilities in Denmark support the utilization of wind energy by establishing a favourable policy in pricing and grid connection. Jointly with the Department of Energy they have taken responsibility for the development of the Nibe mills (see section 4.3) and are engaged in the development of a new WEC with 60 m diameter and 1.5 - 2.0 MW rated power.

Plans are under way for a wind farm of 5x750 kW wind turbines. There will be an updated version of the Nibe-B concept, located at a site called Hasuedso. The purpose of the project is to gain experience from the operation of wind farms with large machines.

France

The French wind energy industry already has a history going back several decades. Eight companies produce approximately 20 different types of WECs with more than 1 kW power. One company is marketing a 100 kW-turbine.

Total sales in 1981 were:

rotor diameter	quantity	total power
< 8 m	200	110 kW
8 - 15 m	16	100 kW
15 m	2	200 kW

The companies involved represent a total manpower of 250 employees, of whom only 50 % work on wind energy.

The national electricity board, EDF, took up initiatives within the new governmental programme and supports wind energy for autoproducers and for territories with high generating costs.

West Germany

The situation reflects little support for small WECs. There are about 8 companies producing approximately 15 different types between 1 and 50 kW; 3 companies represent industrial groups. Two large companies have developed medium-sized WECs between 100 and 370 kW. Two large companies have developed multi-megawatt turbines.

The interest in small WECs is increasing as a result of the Danish successes and there are indications that medium-sized companies will participate in development and production.

The large companies are not yet pushing forward in the field of medium and large WECs as they want to gain operational experience before marketing the units.

As regards the production of components, the high industrial standards provide a global capability. In the case of large WECs, German industry has already delivered gearboxes and large bearings for foreign units.

The involvement of the German utilities has been limited up to now to the provision of regulations for autoproducers and to its active role in large WECs by participating in the operational company for the 3 MW-Growian.

Greece

The renewed interest in wind energy motivated some engineers to form the Winsol company several years ago. Up to now, it is the only Greek company to produce small-WECs. Several units of 3 different types (1 kW/ 2.5 kW/ 5 kW) have already been installed, and the development of WECs with higher power ratings has started.

The national utility, PPC, is taking part in testing 5 small WECs in the grid of the isle of Kythnos which is fed by diesel generators.

Ireland

Of the 12 WECs tested within the national programme, only one is of Irish production: this unit is designed to produce heat by using a water brake. The participation of Irish industry will depend on the results of the running programme.

The national utility, ESB, played an active role by testing 4 turbines and by providing regulations for the connection of small WECs to the grid and for the reimbursement of autoproducers.

Italy

Six companies have established a production series of approximately 10 WECs, and other companies are preparing to start production. Initial design studies are taking place for a 2 ÷ 5 MW-sized unit. The companies involved range from medium-sized companies to large ones. Especially in the test programme the large companies Aeritalia and Fiat are playing a leading role.

The national utility, ENEL, launched activities in wind energy by initiating the Vele-project and sponsoring, in conjunction with ENEA, the development of a large WEC (Gamma Project).

The Netherlands

Since 1980 about 10 Dutch manufacturers have started to produce approximately 20 different types of small and medium WECs. One of these WECs has a power rating of 265 kW. Production is intended for both the home and export markets. In this respect, one contract from the USA must be mentioned - for the delivery of 65 units of a 100 kW turbine to California. On the other hand, about fifteen trading companies import small WECs, mainly from Denmark, but the share is decreasing in favour of national products.

The utility companies, represented by SEP, are positively engaged in various actions the main one being to finance 50 % of the national project to install a wind power station of 10 MW until 1984.

United Kingdom

There are currently about 10 firms manufacturing wind turbines in the United Kingdom. The majority are concerned with small machines either for domestic use or for battery chargers in remote locations. Many of the manufacturers are themselves small businesses and in some cases have sold only a limited number of machines.

At the lower end of market (with respect to machine size) there are several firms offering wind-powered battery charging systems. Rated sizes vary from 5 W to 120 W. Applications include: offshore use by port authorities, oil companies and on yachts, and on land in agriculture and remote telemetry stations for water boards. One firm claims to have sold about 1,000 units in 1982 but the average is much less than this.

There are also a small number of firms making wind generators in the 1 - 5 kW range. These are a mixture of machines including ordinary horizontal axis designs, vertical axis (Musgrove variable geometry) machines, and unconventional contra-rotating sail-wing rotors. Again the number of units sold is generally small, consisting mainly of pre-production units. One manufacturer claims 50 machine-years of operations.

Most of the interest in development and future production is concerned with large machines and a number of important companies and consortia are working within the framework of the UK National Programme.

Designs for both horizontal and vertical axis machines in the MW size are being pursued and there are good prospects that both will reach the prototype stage at least. The two main consortia both contain leading civil and mechanical engineering contractors, firms from the aircraft industry, and electrical engineering consultants.

Apart from these integrated design and manufacturing groups, there are some smaller specialist design teams. One of these has licensed its machines for production in Switzerland, while another has built its own medium-sized machine (100 kW) and is seeking financial and industrial partnerships.

The role of the electricity supply boards is being further motivated by the performance of studies about the wind energy potential and integration problems. The Scottish board, NSHEB, is participating in several joint ventures to evaluate British aerogenerators, the largest one to be a 3 MW turbine on the Orkneys. CEGB has installed a 200 kW machine in Wales (1982) and intends to purchase a proven MW-size machine scheduled for 1985.

5. ECONOMIC ASPECTS

The future utilization of wind energy must be envisaged in the context of the existing energy sources with different characteristics, as regards quality, availability, price, etc. Costs and value may be perceived differently according to the interests of the customer. In order to gain independence a private owner may be prepared to invest more than the cost of traditional energy at the time of investment. The same principles might be applied at the governmental level if political implications arise in the traditional energy supply. Other aspects might be employment versus energy imports, the exportation of a new product, or the reduction of pollution, etc.

"Economy" should not be measured solely in terms of money - nevertheless money is the usual yardstick. Therefore, the discussion of wind power will first explore the costs of the existing converters in line production, and then the allowable costs if operated as new types of power stations.

5.1 Analysis of the Costs of Existing WECs

Every wind energy meeting about includes one or more papers on costs. Despite these publications, prices on a commercial basis are not readily available. The data for of the wind turbines listed in Table 4-1 are based on the relatively firm price statements obtained from three main sources:

- a summary of the Danish small WECs [5-1],
- the IEA report on costing [5-2],
- information provided by the manufacturers.

The price basis is 1981/82.

Because quotations are given for different conditions - with or without transport, taxes, foundation, etc. - all prices have been controlled for comparability and appropriate corrections have been made to cover all expenses, including the grid connection. This procedure involves a certain probability of inserting additional unknowns, e.g. for foundation, site preparation, etc., but it was found that the overall comparability was improved. Furthermore, the figures can be used directly to calculate the energy costs.

It should be noted that great price differences exist - occasionally up to 50 %. As the aim of this study is to show how far the technology has been developed, wind turbines at the lower end of the price range were selected, subject to proven design and reliability. In the diagram covering the total investment costs (Fig. 5-1), all licensed Danish small WECs have been plotted to show the range of scatter.

The energy price depends on the capital costs and operating costs. As the consumption of fuel, lubricants, etc. is negligible for WECs, the main impact on operating costs will come from maintenance and repair, if automatic operation is assumed. Wind energy utilization is at too early a stage to provide exact figures. However, in chapter 4.1, a Danish investigation was cited [4-2] which gave annual operational expenses as 1 - 2 % of the investment, based on 120,000 operational hours of 32 WECs (30 and 55 kW).

A theoretical analysis of large WECs [5-3] indicated annual M&R expenses in the order of 1 % of investment, based on experience with wide-body aircraft.

Another analysis of large WECs, based on experience of the Nibe mills in Denmark, shows annual operational expenses of less than 1.5 % of the initial investment. (This information comes from Elsam, the power company responsible for the operation at Nibe.)

Only the experience which will be gained from the increasing number of WECs will determine the real expenses; nevertheless, values below 3 % can be expected. It is also well known that the need for maintenance and repair can be reduced by adequate design and production.

Despite the preliminary nature of the cost evaluation, the general tendencies derived from converters from

- Denmark
- West Germany
- The Netherlands
- Sweden
- United States

permit first-order cost predictions.

Figs. 5-1, 5-2 and 5-3 show the parametric cost comparisons. In many cases, the total costs, including installation and grid connections, will be of interest, as in Fig. 5-1. Three main design principles were identified:

- small WECs with fixed rotor blades, the rotational speed being controlled by a strong asynchronous generator and at high wind speed by aerodynamic stall;
- small WECs with fixed rotor blades, the rotational speed being free since an electric AC-DC-AC inverter provides the correct voltage and frequency;
- small and large WECs with pitch control of the blades.

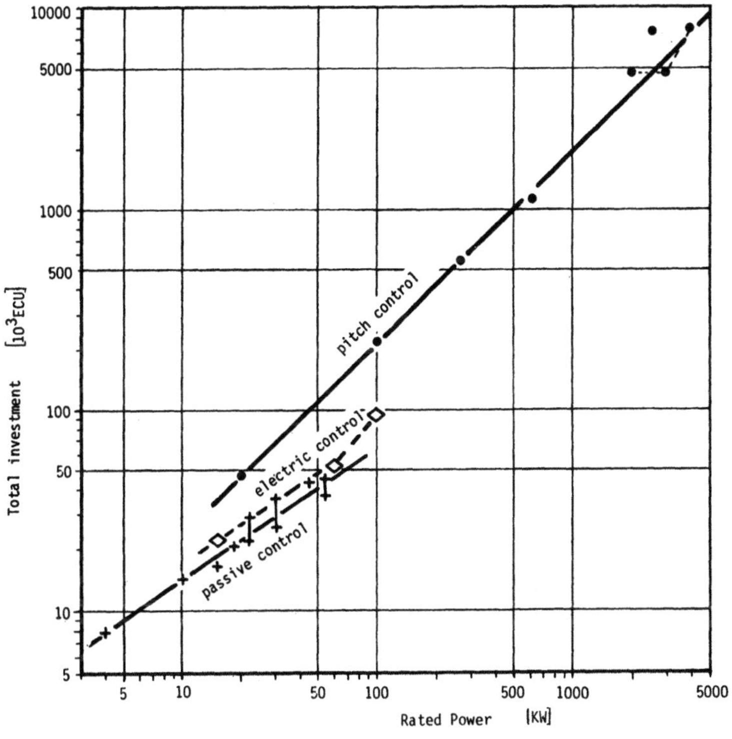

Fig. 5-1:
Comparison of the Total Investment Costs for Various WECs

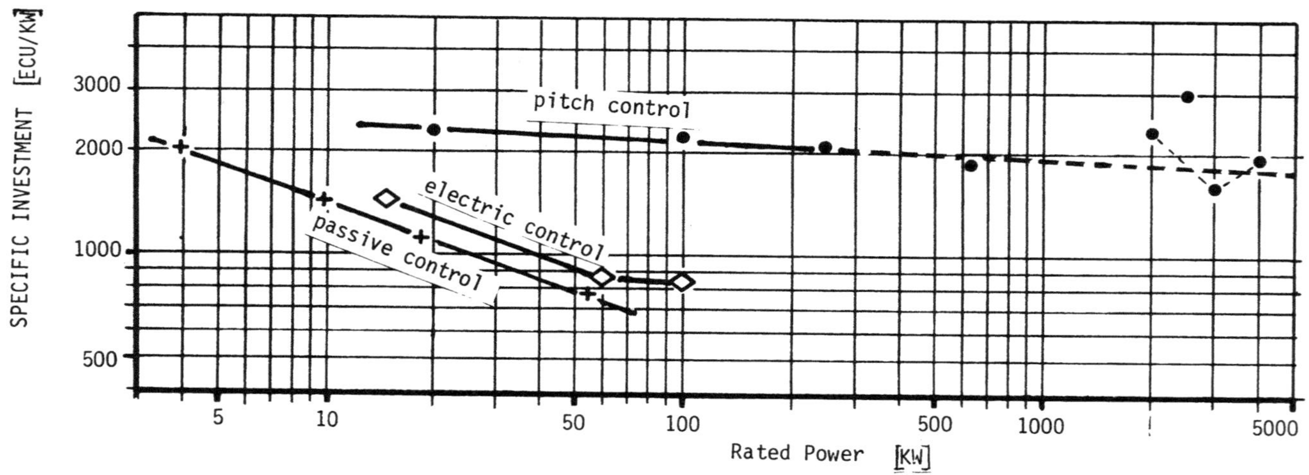

Comparison of specific investment costs of various WECs

Fig. 5-2:
Specific Investment Costs versus Rated Power

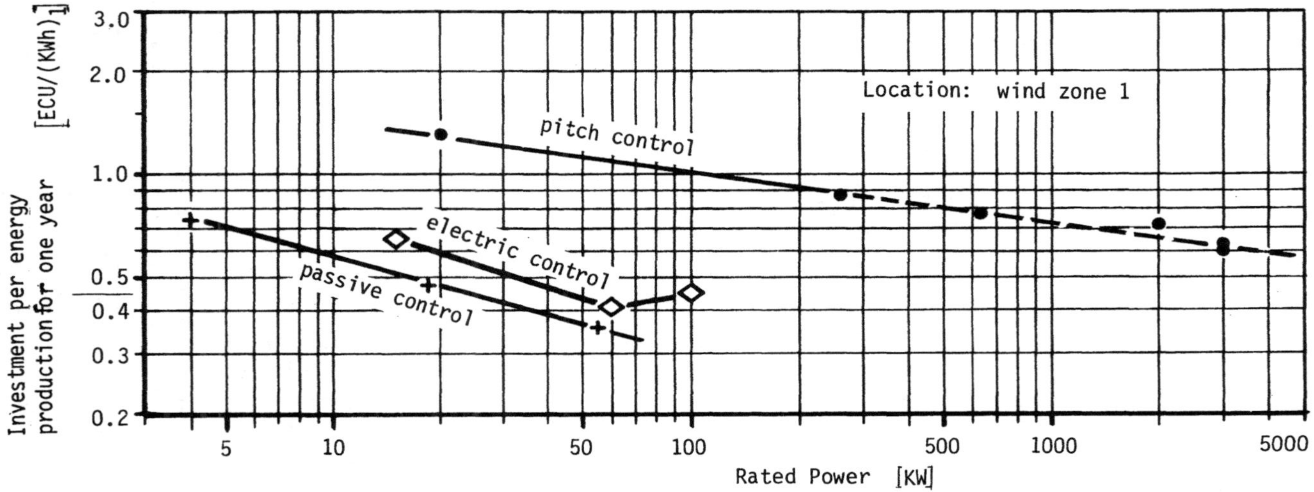

Investment of energy costs, if the total investment is allocated on the energy of one year

Fig. 5-3:
Comparison of Energy Costs if the Total Investment is Allocated to the Energy Produced in One Year

Pitch control results in a more complex design and thus raises costs. Electric control is more expensive than passive control, but the cost differences are comparable to the price bands of various suppliers.

It is surprising to find such a good correlation over four orders of magnitude. The deviations due to additional charges and individual offers may range in the order of 30 %. Another influence stems from the rather free selection of rated power. No major cost increase is involved in selecting a larger generator and thereby improving the data. For this reason, price data per installed power (Fig. 5-2) must be read with care.

The power/cost relationship bears out the technical experience that an increase in size results in a decrease of specific costs. For small WECs, the lower price level was found at about 800 ECU/kW, for large WECs at about 1,800 ECU/kW.

It must be borne in minde that small WECs represent a second or even third generation of rotors, already in line production of 10 to 100 units per batch. The large WECs are prototypes, with price expectations given for a 10 unit production.

The specific investments are of interest for comparative calculations with respect to conventional power plants. The second value needed, the "equivalent utilization period at maximum output capacity", is obtained by dividing the total annual energy (Fig. 4-4) by the rated power.

In most cases, the energy costs provide the best comparison. As escalation rates, interest rates, commercial and technical lifetimes complicate the picture and open up the possibility of selecting advantageous assumptions, a neutral basis of comparison was found: i.e. the total investment was divided by the total energy produced in one year, expressed as $ECU/(kWh)_1$, the index 1 being a reminder of this special definition. Division of this figure by the lifetime gives the investment per actual kWh without charges for interest, M&R costs, etc. As the $ECU/(kWh)_1$ figure is the capital needed to obtain a supply of 1 kWh for n years, the annualized capital cost (ACC) assuming constant annual amortization, can be calculated using the conventional formula:

$$ACC = I \cdot \frac{i}{1 - (1+i)^{-n}} = I \cdot R$$

where I = investment
i = interest rate
n = lifetime in years

The capital recovery factor R will be for example:

	n	R
i = 0.05	20 30	0.0802 0.0651
i = 0.10	20 30	0.1174 0.1061

In wind zone 1, the lowest energy costs are obtained for WECs of 50 - 60 kW size, with 0.36 - 0.40 $ECU/(kWh)_1$. The actual costs per kWh will then be

capital costs:	0.36 x 0.08	0.40 x 0.08	5 % interest, 20 years
M&R costs:	0.36 x 0.02	0.40 x 0.02	2 % of investment for
	0.036 ECU/kWh ÷	0.040 ECU/kWh	maintenance and repair

For pitch-controlled large turbines, comparable energy price values of 0.5 $ECU/(kWh)_1$ are indicated. The actual costs per kWh is then:

capital costs:	0.6 x 0.065	0.7 x 0.065	5 % interest, 30 years
M&R costs:	0.6 x 0.02	0.7 x 0.02	2 % of investment for
	0.051 ECU/kWh ÷	0.060 ECU/kWh	maintenance and repair

The price gap between the most economical small WECs and the first generation of large WECs is still noticeable, as is the significant influence of for maintenance and repair costs. It is assumed that small WECs will have a 20-year lifetime and large WECs 30 years.

The graphs do not explain the existence of the very large cost differences between small and large rotors. The last plot showed that the actual energy costs might not differ significantly because of the different lifetimes. The difference is however, accounted for mainly by the simplicity of the design of small rotors, coupled with the benefit of series production. This design principle (fixed rotor blades) may be extended a little further, but cannot be used for large rotors owing to the stresses and safety factors - so far as is known today. Change of the design principle is accompanied by a jump in costs which is however susceptible to reduction by size enlargement. Moreover, simplification and series production can be expected to lower the costs of large rotors.

A question mark must be put against medium-sized WECs, which suffer by not using simple technology but cannot profit from size benefits. It should be noted that this comparison covers only a few aspects, albeit the most obvious ones. Further - unquantifiable - influences may come from actual land use considerations, the system requirements of a grid, etc., therefore these graphs cannot give a final answer for the time being.

As mentioned above, systematic ordering of the quite differently based data is eased by converting it to comparable conditions of the same site. It is assumed that the data for annual energy versus mean annual wind speed have been reported correctly. Hence, the general tendency of the lines of energy costs versus wind speeds at hub height (Fig. 5-4) enables an extrapolation to different wind speeds at the same hub height. Especially the range of large rotors, for which we do not yet know many details of performance and costs, will be rendered accessible to cost assessments, although the accuracy is still low. For example, the comparable energy costs at 10.5 m/s may range in the order of 0.35 $ECU/(kWh)_1$ whilst the costs at 6.5 m/s would be approximately 1.05 $ECU/(kWh)_1$, or 3 times more.

5.2 Cost of Wind Energy in Decentralized Applications (private ownership)

This section discusses privately owned WECs which are mainly small units employed for autonomous electricity production.

5.2.1 Energy Costs of Small WECs Connected to the Grid

Most of the information needed here has been presented in the graphs of the preceding section. If the length of the cable connecting the turbine to the consumer and to the grid is not too long, the cost data can be used directly. Normally, the electricity produced cannot be consumed by the owner at all times, and the grid connection enables surplus energy to be sold to the networks, as discussed in Section 5.2.3.

Most of the WECs available on the market need grid a connection to magnetize the asynchronous generator for constant frequency output.

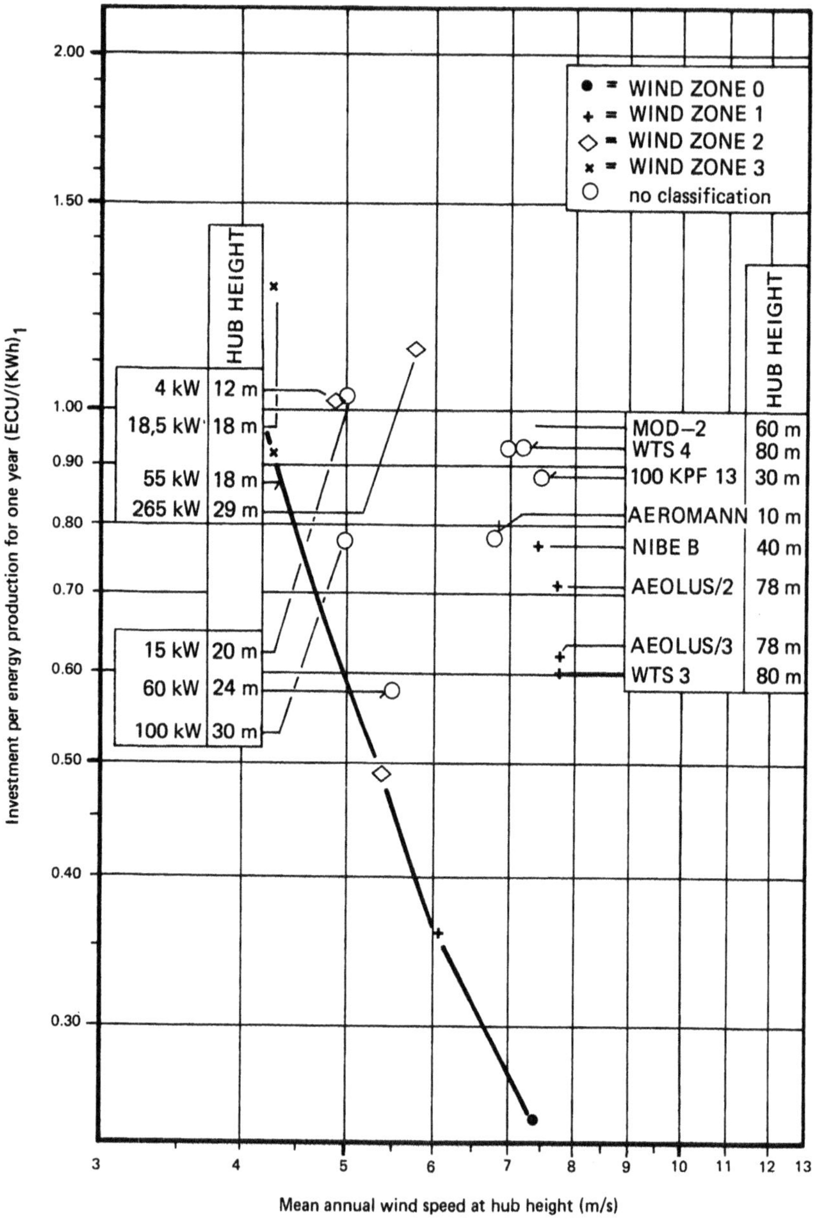

Fig. 5-4:
Energy Costs of Various WECs at Different Wind Speeds

5.2.2 Energy Costs of Small WECs Without Grid Connection

The rotors suited to these plants need active control of the rotational speed by internal electric devices, or pitch control. Hence the availability of WECs is reduced and the prices are higher than for passive control. The autoproducer will need a storage device to ensure that a certain power level is constantly available. The storage device will increase the overall costs significantly and its size should be optimized, especially in connection with the statistical assumption of how many hours of no power at all can be tolerated. The following summary of an analysis by ENEL for Italian conditions brings out the main influences.

ENEL surveyed 601 isolated houses that might have a very high interconnection cost owing to their distance from the electricity grid. The results indicate that about 50 % of the sites investigated have a suitable wind source availability.

The possibility was examined of supplying the energy needs of the above houses by means of a simple stand-alone generating plant (WEC + storage) which is shown schematically in Figure 5-5.

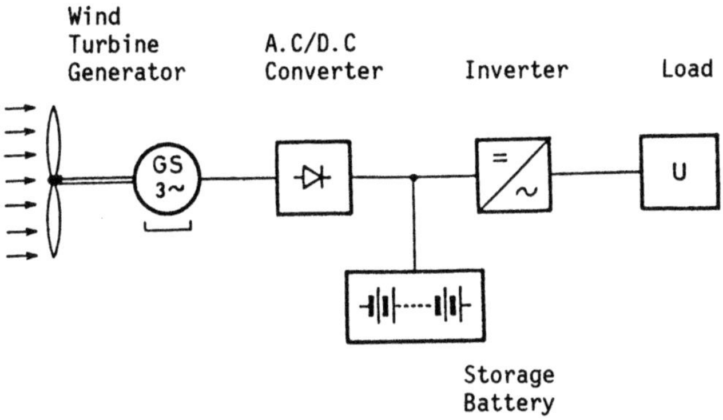

Fig. 5-5:
Electrical Components of a Stand-Alone System

In the study it was assumed that the load to be supplied was of a residential type, characterized by an annual absorbed energy equal to 1,800 kWh and by a yearly peak equal to 2.8 kW.

Evaluation of the maximum generating and storage capacity to be installed was performed by a methodology that, on an hourly basis, compares the amount of wind energy available with the energy demand. The size of the wind generator and of the storage battery is calculated in accordance with the criterion of providing the load with an assigned reliability level.

Moreover, since an assigned supply reliability can be obtained by several combinations of generating power and storage capacity, a unit cost was assumed for both of them in order to determine the size of the plant at minimum capital cost.

Figure 5-6 shows the design curves of the wind generator + storage for three different reliability levels (i.e. annual number of hours when the load is not supplied) and for two different cost assumptions.

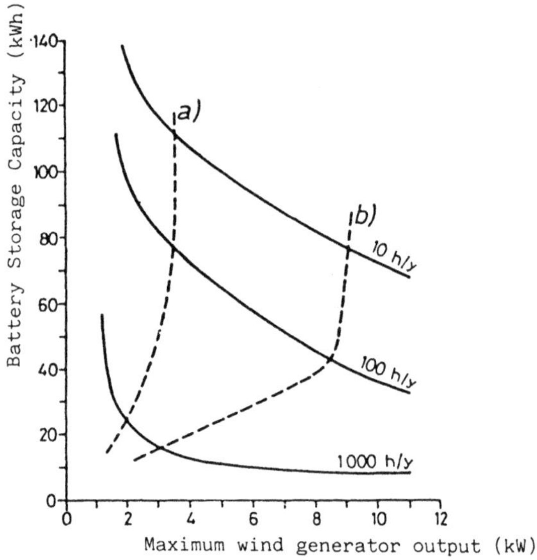

Wind generators

	ASSUMPTION (a)	ASSUMPTION (b)
Capital cost ($ECU/_{kW}$)	3030	1136
lifetime	20 years	

Storage battery

	ASSUMPTION (a)	ASSUMPTION (b)
Capital cost ($ECU/_{kW}$)	90.91	75.76
lifetime	5 years	

Interest rate 5 %
(1 ECU = 1300 It.L.)

Fig. 5-6:
Optimization of WEC-Battery Systems

By means of the design curves of Fig. 5-6, the size of the wind generator and of the storage battery giving the minimum capital cost of the plant, as well as the corresponding unitary cost of the energy delivered to the load, have been evaluated and are shown in the following table.

yearly hours of load supply unavailability	Cost assumption "A"			Cost assumption "B"		
(hours)	Maximum wind genera tor output (kW)	Storage capacity (kWh)	Energy cost (ECU/ kWh)	Maximum wind genera-tor output (kW)	Storage capacity (kWh)	Energy cost (ECU/ kWh)
10	3.5	112	1.8	9	77	1.23
100	3.5	77	1.4	8.5	43	0.87
1000	2	23.5	0.6	3	17	0.36

With reference to the energy cost found for the case of a reliability level of 100 hours/year of load supply unavailability, the following evaluations were made:

- Based on cost assumption "A", wind energy is more economical than grid interconnection in the case of 30 % of the sites where wind energy utilization is possible,
- Based on cost assumption "B" (which may be considered a long-term project cost), the above percentage rises to 50 %.

Using the information from the preceding graphs of the assessment study, the analysis without storage capacity is as follows:

- Fig. 4-4 shows the annual energy of a 4 kW-rotor to be approximately 7,500 kWh per year;
- Fig. 5-2 indicates 2,000 ECU/kW as the lowest investment, the total investment being 8,000 ECU;
- as only 1,800 kWh/a are needed,the investment per annual kWh will be 4.44 $ECU/(kWh)_1$. This corresponds to an annualized capital cost of 0.37 ECU/kWh consumed if 5 % interest rate is assumed.
- As a rough estimate, an outage (i.e. no power at all) of 3000 h must be expected.

Comparison of the energy costs for the rotor (0.37 ECU/kWh) with the results of cost assumption "A" indicates the high price to be paid for supply certainty.

5.2.3 Tariffs for Autoproducers

It is evident that the timing of energy production by a WEC and the timing of consumption will differ to a certain degree and that the utilization of surplus energy will bring an economic benefit if it is paid for on an appropriate scale.

The aim of the Electricity Generating Boards is to produce electricity, and therefore their response to the new situation - the purchase of energy from private owners of WECs - has been initiated but not yet finalized.

According to the degree of public interest and the progress of wind energy technology, the tariff regulations now in force in the various countries range from non-existent to favourable conditions for wind power. The most advanced conditions have been adopted in Denmark and they will be reported in some detail below.

Appendix 6 describes the situation in the other countries of EUROPE-10.

Denmark

The accounting rules described below apply to renewable energy sources, including wind turbines. The organization of the energy plant determines the accounting method. There are 3 possible cases:

1. Individually owned wind turbines.
2. Jointly owned wind turbines using a private network for distribution.
3. Jointly owned wind turbines using the public grid for distribution.

In cases 1 and 2, the energy supplied to the owners from the public network and the owners' surplus energy generated by the wind turbine are measured separately by two kWh meters. The first meter measures the energy flow in one direction, the second in the other direction.

In case 3, the energy generated by the wind turbine is measured by one kWh meter. The consumption by each owner is measured by the normal kWh meters.

Electricity supplied from the public grid to the owner/owners of the wind turbine is billed according to the regular tariff. If the grid connection network has been reinforced, the owners are billed according to the extra investments.

When electricity is sold to the power distribution company, the rules for payment are as follows:

Cases 1 and 2. The amount paid by the power distribution company is made up of a capacity charge and an energy unit charge. In principle, the payment corresponds to the savings made by the power distribution company.

- o The energy charge is identical to the energy charge that the power distribution company pays to the power generation company plus 10 per cent to cover grid losses.
- o The capacity charge is included in the kWh price, the amount corresponding to 50 % of the capacity charge that the distribution company pays to the generating company.

Case 3. The accounting of electricity sold to the power distribution company follows the same rules. However, some additional rules are needed to deal with the electricity generated by a jointly owned wind turbine. First of all, two conditions must be fulfilled when the public grid is used for the distribution of "private" energy.

- o The joint owners and the wind turbine must be connected to the same power distribution company.
- o The distance from the wind turbine to any owner may not exceed 3 km.

To determine how much of each owner's consumption is supplied by the wind turbine, the following approximation is used:

- o The wind turbine is assumed to supply a maximum of 60 per cent of its production directly to the owners.
- o The owners are assumed to receive a maximum of 55 per cent of their annual consumption from the wind turbine.

Both sets of conditions must be fulfilled.

At the end of the accounting period, the readings of the individual kWh meters are added together to find the total consumption by the owners. The wind energy production is read from a separate kWh meter.

The electricity purchased from the power distribution company by each individual owner is calculated as the difference between the total consumption by this owner and a proportional quantity of electricity supplied by the wind turbine.

The sale from the wind turbine to the power distribution company is calculated as the difference between a proportional quantity of the total wind turbine production and the electricity supplied by the wind turbine.

The above accounting rules for wind turbines are provisional, and revised edition is in preparation.

5.2.4 Public Financial Aids to Wind Energy

Almost all countries have research and development programmes in renewable energy. At the marketing phase of the new devices, the differences in the present national ratings of these technologies become apparent. It is well known that wind energy utilization is booming in the USA owing to about 20 Federal and State acts providing direct and indirect financial support. The main acts are

- o The Crude Oil Windfall Profits Tax Act provides for a 40 % Energy Tax Credit (ETC) for residential solar and wind powered equipment (with a $ 4,000 limit on a $ 10,000 system). The Act provides business ETCs for wind installations of 15 %.
- o The Economic Recovery Tax Act (ERTA) makes sweeping changes to the U.S. Tax Code. These changes are intended to provide a general stimulus to the economy by encouraging investment of all forms.

Major changes in the tax laws include the following:

- o Tax Rates:
  In addition to the general reduction in tax rates for individuals and corporations, maximum tax brackets are reduced from 70 % and 50 %. This reduces the after tax value of many tax shelters while increasing the attractiveness of profit making ventures.
- o Depreciation:
  The complex Asset Depreciation Range (ADR) system is changed to an Accelerated Cost Recovery System (ACRS). Personal property is divided into four classes: 3 years, 5 years, 10 years and 15 years public utility property.
- o The State Solar Tax Credit (expired end of 1983). The State provided a maximum tax credit for 55 % of a system's cost with a limit of $ 3,000 for any system which cost less than $ 12,000. Federal energy tax credits could be added to the State credit, however, the total could not exceed 55 % of the costs. The State also provided a credit for systems not installed on a dwelling and which had a cost greater than $ 12,000, equal to 25 % of the costs with no limit on the State contribution.

More information is given in reference [5-4].

Similar laws already exist in some European countries. The situation is changing continuously as tax provisions and subsidies are used as instruments to control the national economy. Owing to these circumstances, only a few regulations can be reported here:

Denmark

The development of wind turbines was greatly stimulated by governmental subsidies to Danish customers amounting to 30 % of the purchase price, provided that the unit had been successfully tested and licensed by the Risö Test Centre.

West Germany

There are several incentives and subsidies for renewable energy. § 82A of the German income tax law allows a depreciation of 10 %/a for all investments associated with the use of regenerative energy sources. The "Investitionszulagengesetz" gives an investment subsidy of 7.5 % for the installation of wind power plants, and in addition a subsidy of 20 % if the wind power plant is associated with research and development activities. If the wind power plant is installed in regions of low per capita income, which is relevant for wind power plant installations in East Friesia and most parts of Schleswig-Holstein an additional 8.75 % investment subsidy can be received. It is noteworthy that the full investment (for example 100 units) can be depreciated regardless of the level of the investment subsidy. In addition, investments for wind power plants could be counted as investments for anti-pollution and would then qualify according to § 7b income tax law for a depreciation of 60 % of the investment in the first year and 10 % each in the following 4 years; up to now, no such application is known of

Italy

According to the law, the task of meeting the national electricity demand is vested in ENEL and some of the municipal electricity companies.

Only households not connected to the grid and industrial firms that are able to exploit their own internal energy resources, or set up cogeneration plants, are allowed to produce their own electricity.

Recently, the Italian parliament, by Law No. 308 of 29 May 1982 on "The Containment of Energy Consumption, Development of Renewable Energy Sources and the Operation of Electric Power Stations run on Fuels other than Petroleum", de-restricted the production of electric power by plants supplied from renewable energy sources, provided the power of such plants does not exceed 3,000 kW.

According to the same law, public funds may be employed to provide up to 80 % of the capital expenditure required, not exceeding the total sum of Lit. 15 million per plant, for the installation of plant to produce electric power from renewable sources to supply permanently occupied rural buildings that have no electricity.

Lastly, public funds are granted for demonstration plants and for prototypes powered by renewable energy sources that are economically viable.

The Netherlands

The application of wind energy in The Netherlands is being stimulated by two major incentive schemes, WIR and IPR. These schemes are applied in conjunction with each other. The total amount of the grants depends on the type of investment and the area in which the investment is made. The main points of these two schemes are summarized below.

The law on the Investment Account (WIR)

This is a fiscal arrangement by which the WIR investment bonuses are either offset against tax assessments or are paid out in cash to the extent that WIR bonuses exceed tax liabilities.

Tax deduction or cash payment is effected in the fiscal year after the committment for expenditure is made. The WIR bonuses vary according to the type of investment (building, open-air plant, equipment), they are not subject to taxation, and do not affect the basis for depreciation.

There are two important WIR investment bonuses: the WIR basic bonus which applies to the whole of The Netherlands, and the special regional bonus (WIR-BRT) which is limited to certain specified areas. There is no ceiling or maximum on either of the WIR bonuses, and they apply to both replacements and new investments. (The WIR-BRT for open-air plant is granted for new or expansion investments only.).

The Investment Premium Regulation (IPR)

The IPR allows cash grants on investment in fixed assets (land, buildings, open-air installations and equipment) in certain specified areas. A grant of 25 % (sometimes 15 % or 35 %) of capital expenditure on fixed assets is available, usually up to a maximum investment of DFL 16,000,000. The IPR applies to investments in industrial and "booster" service enterprises, new or extended. The IPR grant is not an automatic right, but is given by the Ministry of Economic Affairs after application and evaluation of a project. The grant is paid shortly after starting up the plant, or it is advanced when one quarter of the investment has been realized and paid. This grant is not considered as income and is therefore not subject to taxation. Depreciation is based on the amount of investment minus the IPR grant.

The European Economic Community (EEC) has ruled that in The Netherlands the net value of all regional grants may not exceed 20 % of the total investment.

This maximum could be reached in areas where both the regional premiums, WIR-BRT and IPR, are payable. (The WIR basic bonus is not considered as a regional premium.) If an IPR grant is given, the WIR bonuses are calculated on the gross investment minus the IPR grant.

Summary of Grants	Date: September 1981	
Grants	Given for investment in	Percentage
WIR basic bonus	Land New buildings Existing buildings Open-air installations Equipment	nil 14 8 12 12
WIR-BRT	Land New buildings Open-air installations Equipment	nil 20 10 nil
IPR	Fixed assets (land, buildings, open-air installations and equipment) - up to DFL 16,000,000 - more than DFL 16,000,000	 15, 25 or 35 Negotiable

Wind Energy

For the application of wind energy there is a third important WIR bonus, namely the energy bonus. For this application a grant of 15 % is available. By cumulating the two grants and the applicable bonuses, incentives up to 40 % of capital expenditure can be reached. Normally they amount to 25 %.

Note:

These regulations are not applicable to institutions exempted from tax, such as the Dutch Electricity Supply Undertakings.

Conclusions:

The above examples show that

- o first, a good R&D programme is needed to provide the scientific and technical basis for wind energy technology, and
- o secondly, a substantial support (in the order of 30 % - 50 %) by governmental/public agencies is essential to motivate manufacturers and consumers in the initial stages, until prices decline owing to series production and the market for WECs developes.

5.3 Allowable Costs of Wind Energy in Centralized Applications

Prospective autoproducers do not always base their decision to own a WEC on commercial aspects alone. In centralized applications of WECs, whether clusters of small WECs or large WECs, the first criterion to be studied will be the economic distance to competing energy generation systems. Other considerations exist, especially at the national interest level, which may outweigh the financial advantages or disadvantages. Nevertheless, if wind energy cannot be shown to meet the needs of the economy, it will be difficult to impose its utilization.

The cost structures of the existing conventional power plants are well known, and the method of estimating the targets for commercial acceptance of wind energy must take these cost structures into account.

5.3.1 Costs of the Conventional Production of Electricity

Despite the easy access to the costs of power plants in particular cases, experts opinions differ on average values owing to the differing conditions on which the average is based.

The following tables are taken from a report of the UNIPEDE-Nuclear Energy Study Committee [5-5], a generating costs assessment made in 1981 for plants to be commissioned in 1990. The energy costs in 1981 prices are summarized in Table 5-1, and the investments presented in Tables 5-2 and 5-3.

10^{-2} ECU per kWh at 1/1/81

COUNTRY	*U.K.*		*F.R.G.*		*Italy*		*Netherlands*		*Belgium*		*France*	
Discount rate	5 %	10 %	5 %	10 %	5 %	10 %	5 %	10 %	5 %	10 %	5 %	10 %
Nuclear plant	2 × 622 MW		1 × 1285 MW		2 × 1000 MW				2 × 1000 MW		2 × 1275 MW	
Investment	2.88	4.79	1.58	2.65	0.99	1.59			1.26	2.07	1.02	1.64
Operation	0.29	0.29	0.47	0.46	0.22	0.22			0.57	0.57	0.36	0.36
Fuel	1.45	1.61	0.82	0.80	0.78	0.83			0.68	0.71	0.69	0.74
TOTAL		6.69		3.91		2.64				3.35		2.74
	4.62		2.87		1.99				2.51		2.07	
Coal-fired plant	3 × 630 MW		2 × 675 MW		4 × 627 MW		1 × 600 MW		2 × 600 MW		2 × 580 MW	
Investment	1.50	2.58	0.79	1.25	0.56	0.85	0.74	1.22	0.59	0.94	0.83	1.35
Operation	0.36	0.35	0.60	0.60	0.19	0.19	0.31	0.31	0.32	0.33	0.29	0.28
Fuel	4.21	4.21	4.02	3.93	2.38	2.38	3.61	3.51	2.59	2.55	2.50	2.48
TOTAL		7.14		5.78		3.42		5.04		3.82		4.11
	6.07		5.41		3.13		4.66		3.50		3.62	
Oil-fired plant											2 × 580 MW	
Investment											0.71	1.16
Operation											0.25	0.25
Fuel											7.03	7.07
TOTAL												8.48
											7.99	
Recall : discount rate used in each country	5 %		(4 %)		5 %		4 %		8.6 %		9 %	

Table 5-1:
Summary of Discounted Average Cost per kWh

Nuclear
Year of commissioning : 1990

Series A
in ECU/kW on 1.1.81

	United Kingdom	*F.R.G.*	*Italy*	*Netherlands*	*Belgium*	*France*
Reference Plant	AGR 2 x 662 MW	PWR 1 x 1.285 MW	PWR 2 x 1.000 MW		PWR 2 x 1,000 MW	PWR 2 x 1,275 MW
Method of Cooling	Sea Water Direct	Atmospheric Cooling Towers	Sea Water Direct		Mixed With Atmospheric Cooling Towers	Atmospheric Cooling
			Discount 5 %			
Construction Costs (1)	1873	1018	682		857	647
Variation in Price During Construction	0	40	0		0	39
Interest During Construction	322	173	83		117	92
Provision for Dismantling	90(•)	27	20		22	33
Total	2285	1258	785		996	811
			Discount 10 %			
Construction Costs (1)	1873	1018	682		857	647
Variation in Price During Construction	0	40	0		0	39
Interest During Construction	702	370	178		251	195
Provision for Dismantling	17(•)	4	3		9	8
Total	2592	1432	863		1117	889

(1) This cost is established excluding dues and taxes. It includes direct costs, indirect costs (industrial architecture, design, site costs etc.), the costs of the owner (general preliminary design work, accommodation, pre-operating costs) as well as spare parts. It does not include the cost of the first fuel charge.

(•) Site restored to its original state.

Table 5-2:
Capital Cost Referred to Net Electrical kW by Nuclear Plants

Coal-fired

Year of commissioning : 1990

Series A

in ECU/kW on 1.1.81

	United Kingdom	*F.R.G.*	*Italy*	*Netherlands*	*Belgium*	*France*
Reference Plant	3 x 630 MW	2 x 675 MW	4 x 640 MW	1 x 600 MW	2 x 600 MW,	2 x 580 MW
Method of Cooling	Atmospheric Cooling Towers	Atmospheric Cooling Towers	Sea Water Direct	Sea Water Direct	Mixed Circuit With Cooling Towers	Sea Water Direct
Desulphurization	Not necessary because of types of coal burnt	Imposed by law, included in cost		Included in cost		Not necessary because of types of coal burnt
			Discount 5 %			
Construction Costs (1)	1014	551	423	520	436	574
Variation in Price During Construction	0	15	0	0	0	18
Interest During Construction	175	50	17	67	33	63
Provision for Dismantling	0	7	0	0	2	0
Total	1189	623	440	587	471	655
			Discount 10 %			
Construction Costs (1)	1014	551	423	520	436	574
Variation in Price During Construction	0	15	0	0	0	18
Interest During Construction	380	104	36	141	68	139
Provision for Dismantling	0	4	0	0	1	0
Total	1394	674	459	661	505	731

(1) This cost excludes dues and taxes. It includes direct costs, indirect costs (industrial architecture, design, site costs etc.), the costs of the owner (general preliminary design work, accommodation, pre-operating costs) as well as spare parts.

Table 5-3: Capital Cost Referred to Net Electrical kW by Coal-Fired Plants

In the calculations, the following exchange rates have been used:

Amount in National Currency for 1 ECU on 1.1.81

Exchange Rate for Purchasing Power Parity (1)	Official EEC Exchange Rate on 2.1.81 (From Section 1) (2)
41.16 BF	41.38 BF
2.507 DM	2.57 DM
0.577 £	0.546 £
5.793 FF	5.95 FF
990.0 Lire	1217.6 Lire
2.788 Guilder	2.79 Guilder
1.04 US $	1.299 US $

As the data for Denmark, Greece and Ireland are missing from the analysis, the contractors of Action 3 supplied general figures, but it must be borne in mind that they are not based on the same assumptions as those reported above.

In Denmark, the costs of conventional power stations (in 1980 prices) are as follows:

	300 MW	600 MW	
Range of price per kW installed. (kr/kW net)	Established Site 2500 - 2900	Established Site 2000 - 2350	New Site 2450 - 2800
Operating cost at 5000 hours per year utilization. (Øre/kWh)	1.8	1.4	

The figures refer to new installations, coal/oil-fired condensing units with sea-water cooling. Included in the price per kW are external facilities for fuel handling, cooling water, etc. Excluded is interest during construction and erection. The operating costs do not include fuel and desulphurization costs.

The Greek power station costs are listed in Tables 5-4 and 5-5.

In Ireland, oil is the main fuel for power stations, supplemented by gas and peat. Some details of the generating plants provided by ESB for 1981/82are listed in Table 5-6. Average values for the fuel costs to be applied in this analysis may be taken as:

oil costs:	0.032 £/kWh:	0.046 ECU/kWh
gas costs:	0.015 £/kWh:	0.022 ECU/kWh
peat costs:	0.027 £/kWh:	0.039 ECU/kWh

The prices of electricity for domestic consumers in Europe are listed in Table 5-7, corresponding to the different annual consumptions defined in Table 5-8 and taken from the report [5-6].

In Denmark 1980, the average electricity price for a low-voltage customer consuming 3000 kWh per year was 54.5 Öre/kWh plus a fixed amount of 255 kr. per year (including VAT and electricity tax), in total 63 Öre/kWh (0.08 ECU/kWh) averaged.

5.3.2 Calculation Methods for Fuel Credit and Capacity Credit

The introduction of WECs into an existing generating system will produce savings in the operating costs of the conventional plants, mainly by savings in fuel. These savings are generally referred to as an "energy credit".

In addition, the introduction of wind energy will produce an improvement in the overall reliability of the generating system, or will cause reductions in conventional system capacity in cases of planned extensions. These types of savings are termed "capacity credit".

As already discussed in Chapter 3.3, these credits are very sensitive to the technical and economic structure of the specific power plant system.

TYPE OF FUEL	POWER PLANT	POWER IN MW GROSS	NET	COST DRA/kW (NET)	NOTES
Lignite	Aminteo I,II	2X300	2X276	93.275	Includes the lignite-mine
"	Megalopoli IV	1X300	1X270	82.137	" "
"	Komnina I,II	2X300	2X276	81.800	" "
"	St.Demetrios V	1X300	1X276	54.772	" "
Coal	S.E.	2X350	2X325	40.891	Includes port and yard facilities
Nuclear	-	1X970	1X900		Design for 0,25 seismicity
Fuel Oil	-	1X300	1X285	26.000	
Hydro Power	Aulaki	2X90		40.933	This cost is not comparable to
" "	Sikia	2X85		43.206	that of S.E. stations because the
" "	Mesohora	2X120		18.604	installed H.E. power capacity is
" "	Glistra	2X90		26.017	disproportionately larger than
" "	Piges	2X105		49.186	the produced energy since hydro
" "	Thesauros	2X172,5		47.853	is primarily used to cover peak
" "	Steno Kalar.	4X180		38.889	demands.

Note: The cost estimate does not include the general expenses of the Public Power Corp.

Table 5-4: Cost of Power Plants in Greece (fixed 1981 prices)

Type of Fuel	Power Plant	Energy for Average hydrological Conditions (GWH)	Fixed Operating and Maintenance yearly cost (Dra/kwh)	Variable yearly Operating and Maintenance cost (Dra/kwh)	Variable Fuel cost (Dra/kwh)	Projected lignite deposits in 1986 (million tons)
Lignite	Aminteo I,II	-	2724	0,055	0,31	270
"	Megalopoli IV	-	3293	0,138	0,28	370
"	Komnina I,II	-	2543	0,055	0,26	106
"	St.Demetrios V	-	1637	0,055	0,15	1305
Coal	S.E. Plant	-	550	0,097	1,376 (2)	-
Nuclear	-	-	861	0,038	0,364	-
Heavy Oil	-	-	450	0,049	2,309 (1)	-
Diesel	-	-	200	0,083	7,904 (3)	-
Hydro Power	Avlaki	309	300	-	-	-
"	Sikia	315	300	-	-	-
"	Mesohora	155	300	-	-	-
"	Glistra	211	300	-	-	-
"	Piges	190	300	-	-	-
"	Thesauros	362	300	-	-	-
"	Steno Kalar.	642	300	-	-	-

(1) Corresponds to specific consumption of 0,24 kG/kwh and fuel cost of 185 Dollars/ton.
(2) " " " " " 0,378 " " " " " 70 "
(3) " " " " " 0,4 " " " " " 380 "
1 dollar = 52 Dra.

Table 5-5:
Breakdown of Operational Costs of Greek Power Plants (in fixed 1981 prices)

Steam Stations

Station	Location	Plant Particulars					Plant Performance for year ended 31/3/'82			
		Generating Sets		Boilers			Units sent out millions	Plant Load factor %	Fuel cost per unit sent out (pence)	Works cost per unit sent out (pence)
		No.	Total Capacity MW	No.	Total Capacity Tonne/hour	Fuel Type				
Arigna	Co. Roscommon	1	15	1	75	Coal	44.7	36.2	3.680	5.833
Great Island	Co. Wexford	3	240	3	853	Oil	393.0	19.6	3.670	4.541
Marina	Cork City	3	120	4	522	Oil/Gas	1,388.0	77.8	1.584	1.889
		1	85	Combustion Turbine		Gas				
Aghada	Co. Cork	1	270	1	800	Gas	2,347.9	74.2	1.491	1.605
		2	170	Combustion Turbines		Gas				
North Wall		3	-†	—	—	—	—	—	—	—
Poolbeg	Dublin City	2 1	240 270	2 1	760 800	Oil Oil	2,188.9	48.2	3.176	3.453
		2	28	Combustion Turbines		Oil				
Ringsend	Dublin City	6	270	7	1114	Oil	137.4	6.5	5.234	10.872
Tarbert	Co. Kerry	2 2	120 500 620	2 2	473 1600	Oil Oil	1,271.3	24.9	3.537	3.966
Bellacorick	Co. Mayo	2	40	2	191	Milled Peat	102.4	31.4	2.476	4.749
Ferbane	Co. Offaly	4	90	4	430	Milled Peat	273.7	37.7	2.944	4.474
Rhode	Co. Offaly	3	80	3	364	Milled Peat	313.0	48.3	2.744	3.849
Shannonbridge	Co. Offaly	2	80	2	346	Milled Peat	307.4	47.5	2.482	3.399
Lanesboro	Co. Longford	2	60	1 1	100 173	Sod Peat Milled Peat	43.5 164.6	26.9 50.0	3.021 2.057	5.188 2.667
Portarlington	Co. Laoise	3	37.5	3	205	Sod Peat	61.4	20.4	3.062	6.140
Allenwood	Co. Kildare	2	40	4	273	Sod Peat	101.5	31.0	3.146	5.713
Cahirciveen	Co. Kerry	1	5	1	28	Sod Peat	8.3	15.9	5.219	11.731
Gweedore	Co. Donegal	1	5	1	28	Sod Peat	8.0			
Miltown Malbay	Co. Clare	1	5	1	28	Sod Peat	2.9			
Screeb	Co. Galway	1	5	1	28	Sod Peat	5.3			

Table 5-6: Particulars and Performance of Generating Plants in Ireland

Furthermore, operational strategies will also influence the optimum share of the various generating types.

For the time being, no operational experience of combined systems can be cited, therefore all publications deal with future system analysis and may be biased by the authors' background. This leads to somewhat reserved estimates by analysts from power generating companies and to a certain overestimation and simplification by wind energy experts. Agreement exists for the energy credit, some refinement being needed for large penetration rates. Large differences of opinion occur with respect to the capacity credit although, in general, this credit is being granted and only the probable value is under discussion. The question of 15 % or 30 % capacity credit is not crucial to the feasibility of wind energy.

Several studies have been published [5-7] or exist as in-house reports in the companies involved. Although they contain valuable information they are "case studies", not suited to a general assessment within the Action 1 task. This is mainly due to the fusion of actual (specific) boundary conditions with mathematical treatment and the reader interested in this topic is advised to study the literature referred to.

A more general approach to economic problems is found in reports by Nissen [3-3], using the mathematical methods developed for conventional power stations and presenting the results in a nomographic form open to individual choices of the boundary conditions. The results were modified for this study so as to widen the range of input data and simplify the contribution by the capacity credit as this would otherwise require detailed knowledge of the power station mix considered but would not exceed second order importance.

The nomographic charts follow two paths:

- one starts from the fuel costs saved within the lifetime of the WEC; the present real value of these costs is calculated by allowing for interest rates and fuel price escalation;
- the other path starts from the installation costs for conventional power stations and assumes the percentage of capacity credit, approximately 30 % at low penetration rates and 10 % - or even less - at high rates.

The evaluation of the fuel credit is based on the following equations:

1. $$K_A = F \cdot \frac{A}{P_r}$$

where:
- F = specific fuel costs in ECU/kWh_e
- A = annual energy of the WEC in kWh/a
- P_r = rated power of the WEC in kW
- $\frac{A}{P_r}$ = equivalent utilization period at maximum output capacity in h/a
- i = interest rate in %
- n = lifetime of the WEC in years

2. $$K_B = K_A \cdot \frac{(1+i)^n-1}{i(1+i)^n}$$

3. $$K_{Ec} = K_B \cdot \frac{\sum_{j=1}^{n} \left(\frac{1+m}{1+i}\right)^j}{\sum_{j=1}^{n} \frac{1}{(1+i)^j}}$$

- m = fuel price escalation in % per year

Table 5-7:
Electricity Prices for Reference Consumers for Domestic Applications
(per kWh in national currency as at 1 Jan. 1982, including all taxes)

REFERENCE CONSUMERS kWh/year	Belgium	F.R. Germany			France		Ireland		Italy		Luxembourg	Netherlands			United Kingdom			Greece
		West reg.	South reg.	Hamburg	Paris Max.	Paris Min.	Urban	Rural	Nth. Cent.	Sth. Isls.		PNEM	PEN	Rotterdam	South Scot.	North Wales	London	
	Francs	Pfennig			Centimes		Pence		Lire		Francs	Cents			Pence			Drs
A 600	7.57	39.79	44.97	40.70	93.76	86.04	8.877	10.627	77.76	77.17	5.492	35.05	35.90	39.18	9.132	8.857	9.687	5.703
B_1 1200	6.05	30.38	34.69	33.11	69.14	63.44	7.627	8.502	87.22	86.63	4.232	28.56	33.15	33.51	5.631	6.873	7.353	5.703
B_2 1700	5.59	26.98	30.53	29.77	71.89	65.97	7.259	7.877	82.87	82.27	3.861	26.65	32.34	31.84	5.190	6.290	6.667	5.703
C_1 3500	5.04	23.03	26.68	25.69	67.15	61.62	6.806	7.106	131.55	130.96	3.015	24.29	31.34	29.78	4.645	5.570	5.820	5.895
C_2 3500	4.41	23.03	25.52	25.69	56.60	51.94	5.911	6.211	(-)	(-)	4.984	23.40	28.83	28.39	4.221	4.630	5.028	4.761
(1300 night)																		
D 7500	4.12	19.22	22.00	22.20	52.29	47.99	5.790	6.014	122.56	121.96	3.028	21.92	28.29	26.89	3.912	4.277	4.624	5.008
(2500 night)	3.97	(-)	(-)	(-)	48.49	44.50	5.565	5.646	(-)	(-)	2.910	21.05	27.53	25.84	3.631	3.946	4.245	5.100
E_1 13000																		
(5000 night)	3.10	13.60	15.83	13.95	42.01	38.55	4.731	4.812	(-)	(-)	2.105	19.37	24.55	23.23	2.744	2.883	3.034	3.559
E_2 13000																		
(9500 night)	3.86	(-)	(-)	(-)	48.29	44.31	5.494	5.578	(-)	(-)	2.874	20.68	27.24	25.42	3.519	3.816	4.096	5.100
F_1 20000																		
(8000 night)	3.00	12.87	15.09	13.40	41.75	38.31	4.651	4.735	(-)	(-)	2.061	18.99	24.23	22.77	2.623	2.742	2.871	3.512
F_2 20000																		
(15000 night)																		
VAT rate at January 1980	17 %	13 %			17.6 %		0 %		8 %		5 %	18 %			0 %			0 %

Table 5-8: Definition of Reference Consumers for Domestic Applications (Annual Consumption in kWh)

	A	B	C	D	E	F
REFERENCE CONSUMER	600	B 1 1200 / B 2 1700	3500	7500	13000	20000
(of which kWh at night)	0	0	(C1 : 0) (C2 : 1300) (3)	(2500) (3)	(E1 : 5000) (E2 : 9500) (4)	(F1 : 8000) (F2 : 15000) (5)
BANDS of CONSUMPTION	< 1000 kWh	1000 to 2000 kWh	2001 to 5000 kWh	5001 to 10000 kWh	> 10000 kWh	
Possible electrical equipment (1)	Lighting Radio TV Refrigerator Various electrical appliances	Equipment of consumer A + washing machine or dish washer	Equipment of consumer A + washing machine + dish washer + C 1 = electric cooker or C 2 = storage water heater	Equipment of consumer A + washing machine + dish washer + electric cooker + storage water heater	Equipment of consumer D + E 1 = direct heating or E 2 = storage heating	Equipment of consumer D + F 1 = direct heating or F 2 = storage heating
Available capacity	≤1,5 kW	3-4 kW	4-7 kW	6-9 kW	≥9 kW	>9 kW
Type of dwelling	2 rooms+kitchen 50 m2	3 rooms+kitchen 70 m2	4 rooms+kitchen 90 m2	5 rooms+kitchen. 120 m2	Dwelling in a residential block 3 rooms+kitchen 70 m2	Individual dwelling 5 rooms+kitchen 120 m2
	(3) for water heating	(4) of which 2500 for water heating	(5) of which 3000 for water heating			

(1) The composition of this equipment is given as an indication and does not necessarily correspond to the situation in all Community countries.

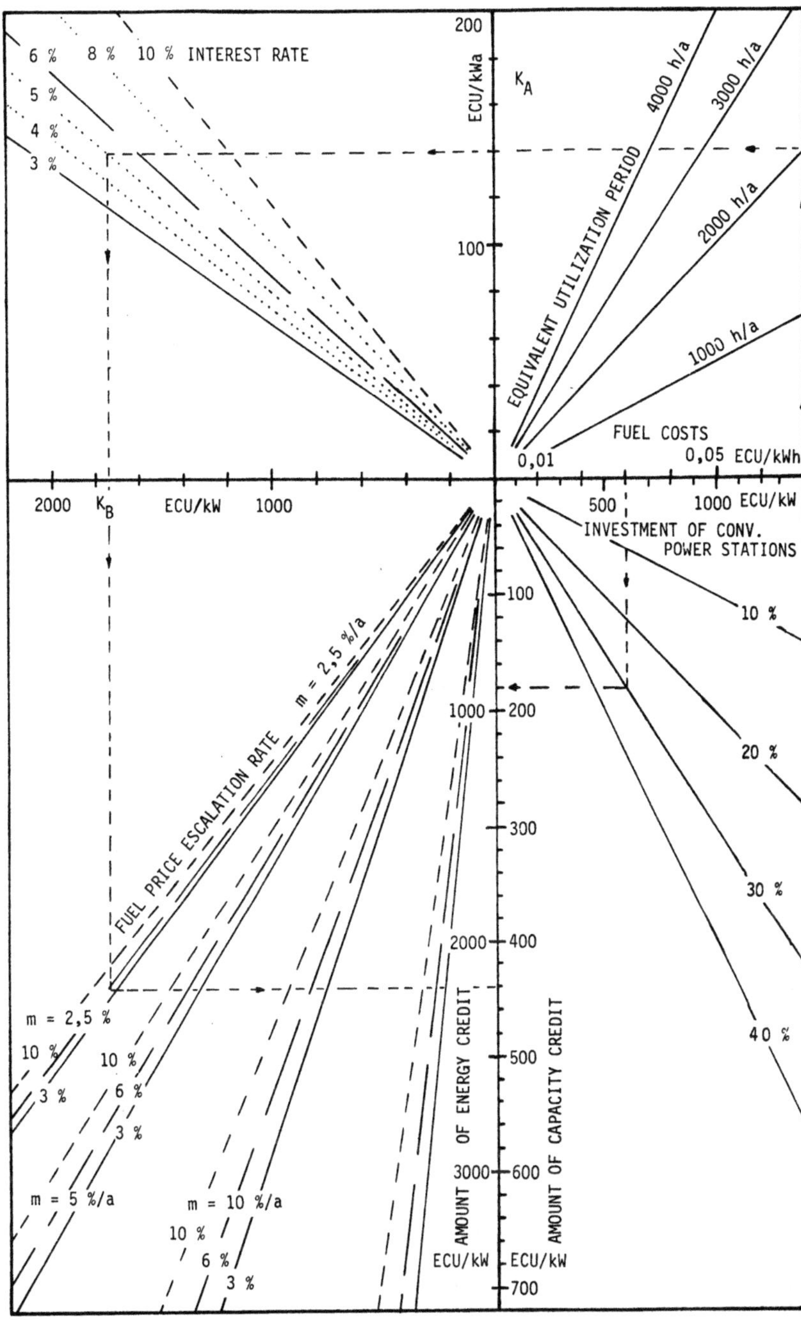

Fig. 5-7:
Nomograph for the Calculation of the Allowable Investment Costs of WECs (lifetime 20 years)

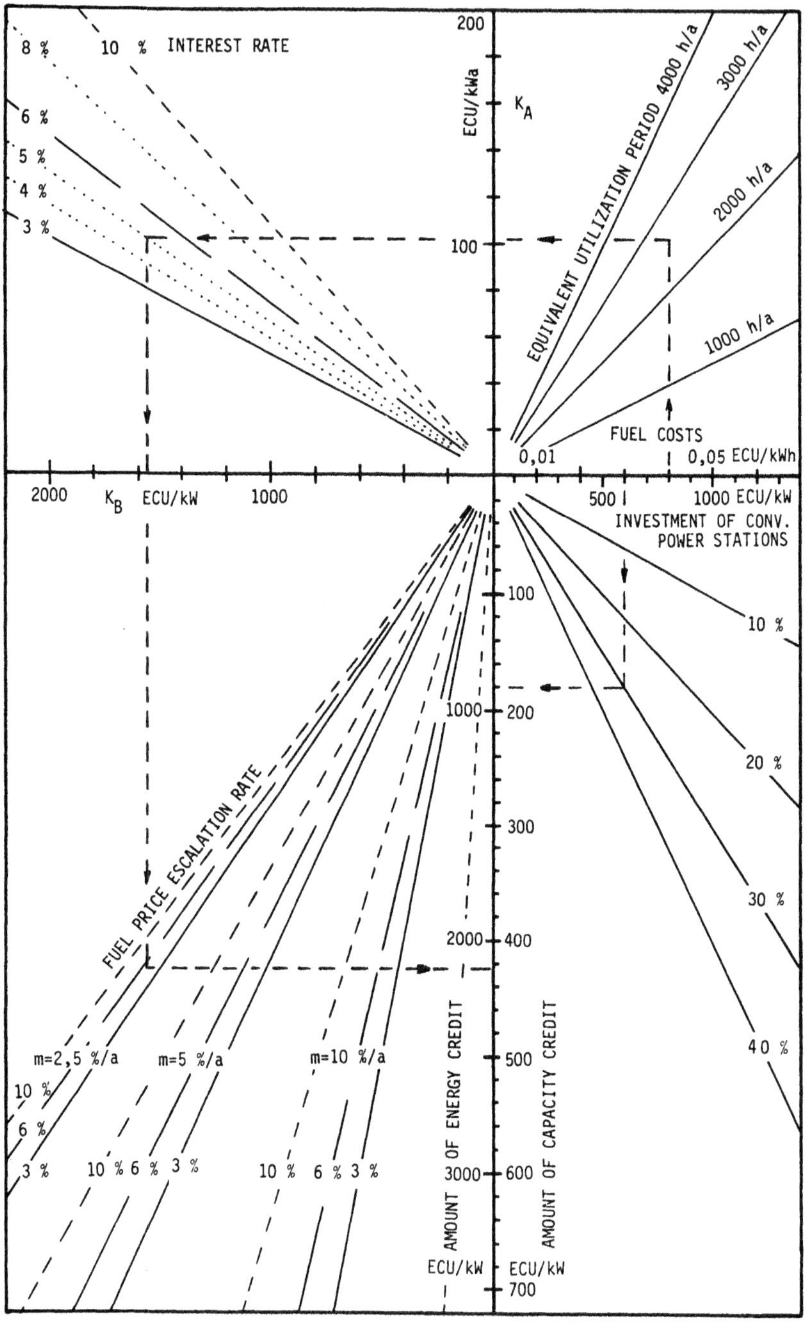

Fig. 5-8:
Nomograph for the Calculation of the Allowable Investment Costs of WECs (lifetime 30 years)

It is recommended to use rates of interest and of fuel price escalation free of inflation. This is intended to keep the calculations in constant currency.

The resulting cost K_{EC} represents the costs avoided by saving fuel, discounted to the monetary values of the year of WEC installation.

The same is true for the cost of capacity credit K_{CC}, but here both power stations are to be discounted, hence the actual investments can be compared directly, neglecting the minor influence of different lifetimes.

The nomographs of Figs. 5-7 and 5-8 have been calculated for 20 and 30 years' lifetime respectively.

The two examples given below are based on investment and fuel costs for oil-fired and coal-fired plants respectively (see Tables 5-1 and 5-3). The equivalent utiliaztion period for maximum output capacity of a WEC was plotted in Fig. 4-4, varying between 2000 and 3000 h/a in areas of wind zone 1. The fuel cost of an oil-fired plant for 2000 h/a at a price of 0.07 ECU/kWh equals 140 ECU/kW per year; when discounted for a 20-year period at 5 % interest rate, this results in 1740 ECU/kW and 2200 ECU/kW respectively assuming a fuel price escalation rate of 2.5 % per year.

The capacity credit would add 180 ECU/kW, assuming 600 ECU/kW of conventional power plant costs and 30 % capacity credit.

The investment for a WEC could be profitable in this example if the specific costs remain below:

K_{EC} = 2200 EUCU/kW
K_{CC} = 180 ECU/kW

Total avoided costs: 2380 ECU/kW.

A comparison with the analysis of existing WECs (see Fig. 5-2) shows that these marginal costs can be met by WECs already available if oil-fired power stations determine the price for electricity. In the case of diesel generators, the cost advantage of small WECs should be even higher.

The second example represents a comparison of coal-fired power plants and large WECs. According to Table 5-1, the fuel costs of coal-fired plants are about 0.04 ECU/kWh for West Germany. For a 3-MW WEC, the equivalent utilization period at maximum output capacity will be 2600 h per year WEC, as shown in Fig. 4-4. This leads to tolerable costs of 2120 ECU/kW if an interest rate of 5 % and a fuel price escalation rate of 2.5 % p.a. are assumed. Again the capacity credit may be 30 %, or 180 ECU/kW.

Thus, in the case of a coal-fired power station, the costs avoided by a 3-MW WEC with a 30-year lifetime sited in wind zone 1 are:

K_{EC} = 2120 ECU/kW
K_{CC} = 180 ECU/kW

Total: 2300 ECU/kW

The question of operational costs is still open. Coal-fired power plants require between 0.002 and 0.006 ECU/kWh in the countries included in Table 5-1. As large WECs are operated automatically, only maintenance and repair need be considered. The amount will be less than 1 % of the installation costs: in the above example, 1 % of 2300 ECU/kW for an annual production of 2600 kWh per installed kW. Hence, an average of 23 ECU must be foreseen for 2600 kWh, resulting in M&R costs of 0.009 ECU/kW. This is more than for operating coal-fired plants, but still in the same order of magnitude, as the national differences might result from different labour costs.

Until more accurate figures for M&R of large WECs become available, it must be borne in mind that some fraction of the allowable cost must be reserved for maintenance and repair. It is hoped that the M&R costs for WECs will drop to values well below 1 % of the investment.

The cost range to be expected for large rotors is 1500 - 2500 ECU/kW in the near future.First priority for replacements should be given to those power plants determining the marginal costs of the systems.

The examples given illustrate use of the nomographic charts for quick reference values but they also indicate the limited accuracy of such a simplified method for a complex cost structure.

The main aspects to be noted are:

- o operational costs: these could be handled as shown above, or neglected in cases of high fuel costs; for the time being, it is probable that the most accurate result would come from not using them at all.
- o completeness of investment costs: attention must be given to both data sources and to comparable cost data;
- o availability: this aspect is included in the data of conventional power stations but is excluded in most WEC quotations; corrections can be made by reducing the equivalent utilization period;
- o plant load factor: some utilities have proved that the large-scale introduction of wind energy might result in a certain percentage of surplus energy which can be dealt with by an appropriate reduction of the permissible investment costs;
- o operational alterations of the existing power plants may occur when wind energy is available, thereby reducing the operational savings or even fuel savings;
- o reserve capacity: this question cannot be covered by these nomograms as it requires a detailed calculation of the power station mix;
- o lifetime: the design goals for most WECs are 20 or 30 years; individual assumptions can be taken into account by interpolation.

Although these aspects may finally cause a shift in the economic results, no change in the order of magnitude is to be expected from them.

5.3.3 Examples of the Allowable Costs for Wind Energy Installations at Specific Locations in the EEC Countries

The data presented in the tables and diagrams can be employed to give a geographical distribution of the costs for wind energy utilization, as well as to indicate the limit of allowable investment costs.

Table 2-3 presented the number of WEC sites and prospective annual energy classified according to wind speeds.

Fig. 4-1 shows that a 100 m diater wind turbine can be assumed to have a rated power of at least 5 MW.

Fig. 4-5 supplies the specific annual energy for the various mean wind speeds at hub height (reference curve) and hence the equivalent utilization period for the various wind speeds at 100 m level (5 MW assumed), as listed in the following table:

$\bar{V}_{100}$	specific annual energy	annual energy for 100 m rotor	equivalent utilization period	percentage of the maximum specific energy
m/s	kWh/m²a	GWh/a	h/a	
> 10	2355	18.5	3700	100 %
9 - 10	2000	15.7	3140	85 %
8 - 9	1700	13.4	2670	72 %
7 - 8	1400	11.0	2200	59 %
6 - 7	1057	8.3	1660	45 %
5 - 6	850	6.7	1340	35 %

Extremely good WEC sites in the class $\bar{V}_{100}$ > 10 m/s were found in West Germany, the United Kingdom, Greece and Italy. The corresponding subfields can be seen in the tables of data compiled for these countries. Ireland is missing here, but this discrepancy will disappear when the Wind Atlas appears.

Assuming fuel costs for coal as listed in the tables, a utilization period of 3700 h/a, an interest rate of 5 %, a fuel price escalation of 2.5 % p.a. and a 30-year lifetime, the following investable cost limits are found by considering the energy credit result alone:

	fuel costs (ECU/kWh)	K_{EC} in (ECU/kW)
West Germany	0.04	3200
Greece	0.026	2050
Italy	0.024	1950
United Kingdom	0.042	3300

Thus without taking the capacity credit into account it may be concluded that wind energy utilization will already be economic at these extremely good sites if the turbines become available at the prices for small quantity production resulting from Fig. 5-2, i.e. approximately 1,800 ECU/kW.

At the sites with low wind speeds of 6 - 7 m/s, the equivalent utilization period will be 1700 h/a, or only 45 % of the highest value. The energy credit will drop accordingly, and it is obvious that the reduced allowable cost level cannot be met by the prospective installation costs of 1,800 ECU/kW unless the above conditions are altered or the capacity credit is added.

The question arises as to how to rate small WECs. Following the method of the data presented, Fig. 5-2 shows the lowest investment costs for 50 - 60 kW wind turbines to be 800 - 900 ECU/kW. These rotors of 15 m diameter run at hub heights of approximately 20 m. At this hub height, the mean annual wind speed will be around 8.0 m/s instead of $\bar{V}_{10}$ = 7 - 8 m/s. According to Fig. 4-5, the annual energy production will be approximately 1150 kWh/m²a, hence 203 MWh/a per 55 kW rotor, or 3690 h/a.

Applying the chart for a 20-year lifetime (Fig. 5-7), the energy credit according to the fuel costs would be (assuming 5 % interest rate and 2.5 % p.a. fuel price escalation):

	fuel costs (ECU/kWh)	K_{EC} (ECU/kW)
West Germany	0.04	2340
Greece	0.026	1530
Italy	0.024	1410
United Kingdom	0.042	2430

Evidently, the existing small WECs will meet these reference investment costs. But care must be taken to allow for additional costs as the intense land use might raise the total installation costs, and the choice of unreliable units might consume the economic advantages through maintenance and repair.

Bearing this in mind, the equivalent values for the lower wind classes will be derived. According to the specific annual energy plot and the wind speed conversion to 20 m height, the following approximate values are found for a 55 kW rotor:

$\bar{V}_{10}$ (m/s)	specific annual energy (kWh/m^2a)	equivalent utilization period (h/a)	percentage of the maximum specific energy
7 - 8	1150	3690	100
6 - 7	850	2730	74
5 - 6	650	2090	57
4 - 5	450	1440	39

If the percentages in column 4 are applied to the energy credits of the preceding table, K_{EC} will drop for the various wind classes. Using Italy again as an example, the 4-5 m/s regime will offer allowable investments by K_{EC} = 550 ECU/kW plus capacity credits of K_{CC} = 150 ECU/kW, hence in total approximately 700 ECU/kW, to be compared with the installation costs of 800 - 900 ECU/kW. Hence, without changing the assumptions (interest rates, fuel costs, fuel price escalation), wind energy will not be compatible at this low wind regime.

5.4 Discussion

There can be no doubt that attractively low prices for small WECs have already been achieved owing to line production by some companies, and the cost analysis underlines the economic sense of their utilization.

An important problem still awaiting solution arises from the optimum size of small WECs. The most economic units (50 - 60 kW) do not really interest the Electricity Boards in Europe because their energy share is too small and for private use these units are far too large. Thus the economy of these machines will be ruined unless the surplus energy can be refunded at a reasonable rate as in Denmark. Since WECs smaller than 50 kW show rather high specific investment costs, their introduction on a commercial basis can be justified only in sites with extreme conditions, i.e. high winds and high fuel costs.

Any further decrease in installation costs is expected to be modest because, despite the large quantities, practical experience is limited to a few years only, and the mainly small companies involved have found it necessary to build up marketing organizations which incur additional costs. The logical line to follow will be to influence the utilities to pay a reasonable rate for the surplus energy from small WECs, as practised in the USA and Denmark.

For large WECs the situation is quite different. The cost projection does not yet exhibit the low energy production costs associated with 50 kW-units, but the large WECs will be operated within the energy supply system, probably by the utility companies, hence the electricity produced will be rated by the same commercial method applied to other power stations. This positive effect will coincide with advances in technology.

At present, the experience with large WECs comes mainly from pilot plants and single units operated under rather special conditions. Only when commercial units have been installed in meaningful numbers,and have been operated for a considerable number of years, will time reveal to what extent the assumptions are correct. An economic evaluation is all the more difficult because the revenue from this application i.e. the money value of the savings on fossil fuels in the future, is perhaps even more uncertain than the expectations concerning cost development. In the course of the cost analysis it was again noted that such an assessment study can only indicate general tendencies with a rather broad tolerance band and will fail if applied to specific cases owing to the large number of variables.

There is no doubt that a large number of sites exist in Europe with a very high wind energy potential. At these sites, the utilization of wind energy would be competitive with fossil-fuel-fired plants even at the production cost level for a batch of 10 WECs, subject to maturity of the units.

A general problem is inherent in the meaning of "economics". A technical professional would expect a very precise definition resulting in an equation. As published by Jäger [5-9] there are at least seven different methods in use which produce a variety of results. As wind energy utilization is associated with long-term investments, a dynamic method was selected for the assessment study; this method is, in principle, used by most Power Generating Boards, although in different forms.

Special care was taken to present the data and methods in a manner which permits the reader to apply personal crosschecks according to this individual assumptions of boundary conditions.

The quantitative effect of storage capacity could not be dealt with. The economic and technical implications of energy storage, in an integrated power system with a substantial wind power component, must be assessed in relation to the system as a whole. Economic evaluations of storage schemes are made in terms of their potential contribution to total generation capacity and possible saving of fuel. However, the operational deployment of storage covers a variety of functions including emergency and scheduling reserve, system frequency regulation, meeting demand peaks and economic generation.

The economic deployment of storage plant for peak-lopping and "in merit" generation is dependent upon the differential cost of generation (i.e. as between periods of high and low demand) being sufficient to cover the cost of the losses in the storage cycle. The addition of a substantial wind capacity would increase the amount of low-cost generation during nights and weekends and thus affect the economics of storage. Savings due to storage would occur if, in periods of low demand, storage were charged, thus eliminating the need either to shut down inflexible plant or to "spill" WEC energy.

In view of the wide ranges in the economic performance of wind energy, and taking into account the uncertainty of most of the input factors leading to these ranges, it is concluded that no statement in this field is possible, at least for the time being. As a general rule, it can be stated that, if the unpredictable development of fossil fuel prices is excepted, the capital cost of WECs is the determining factor in the feasibility of wind energy.

There are many shortcomings in a comparison of wind energy converters and the present well developed power stations. As a consequence of the normal introductory phases, large capacities from WECs should instead be envisaged in combination and competition with future types of power sources and distribution:

- o advanced coal power stations with non-polluting and highly efficient combustion systems,
- o advanced nuclear reactors with complete recycling and waste disposal infrastructure,
- o photovoltaic power plants,
- o large-scale co-generation, e.g. combined electricity generation with long distance heat supply,
- o large-scale production and use of hydrogen.

Present cost comparisons can serve only as indicators of future economic viability.

6. MAIN RESTRICTIONS ON WIND ENERGY UTILIZATION

The approach chosen for the assessment study has favoured and developed technical arguments. There is consequently a danger that an unduly positive impression may be given which could be misleading in reality, as many decisions are based on irrational opinions. It is therefore advisable to discuss some of the as restrictions on wind energy utilization. To estimate the real consequence of these - in most cases indirect - influences is impossible, but the importance of the non-quantifiable amplifying or damping elements should be stressed.

6.1 Legal Constraints

The situation is characterized by the introduction of a new technology into an existing structure of laws, rules and specifications for conventional products. No legal statement exists within the EEC countries forbidding wind turbines explicitly. But many laws and regulations may be - or may not be - applied in case of an installation permit. The discussion starts with the question of whether to treat a wind generator system as a building or as a machine. As the local authorities are responsible for the safety of the general public, the necessity of a building permit in all cases where there could be a possible public danger cannot be disputed, yet the uncertainty as to the correct evaluation of the hazards and the appropriate classification within the code of regulations results in a wide range of personally based decisions, from no objection at all to strict prohibition. In several cases the diverging interests have had to go before the courts.

To overcome these problems, a legal basis which will describe the WEC explicitly is needed. Such rules already exist in Denmark and - since 30 March 1982 - also in the State of Schleswig Holstein in West Germany.

Several areas of public interest are subject to legal control:

- o building permits,
- o safety,
- o protection against annoyance,
- o non-disturbance of traffic, especially airport clearance,
- o non-disturbance of radio and microwave communications,
- o preservation of land areas in the public interest, etc.

In the cases of traffic and electromagnetic wave systems, the authorities have established precise rules about the restrictions on building. These "clearance zones" are well defined, they have been published, and attention can be paid to them when sites are being investigated.

Another source of restrictions based on the legal provisions stems from the monopoly or quasi-monopoly of the utility companies. Today, private ownership of WECs is allowed in all EEC countries, but the sensitive point is the connection to the grid. The integration of wind energy into the public electricity supply system is apparently a restrictive factor of major importance. The maximum wind power capacity which can be integrated into the conventional electricity production system is on the one hand dependent on the predictability of the wind supply and on the other hand on the operating flexibility of the conventional production park and the requirements formulated with respect to the reliability of the electricity supply.

In order to maintain quality standards, the utilities demand rather precise voltage and frequency values for the power to be fed into the grid. No technical problem is involved in meeting these requirements, but the telated expenses influence the economics of the WEC.

This leads to another restriction in practice, namely repayment of the energy fed into the grid. Most utilities claim that they gain no benefit at all from the unreliable surplus energy of small WECs and hence they pay only for the fuel saved on a commercial basis. Yet the economy of small

WECs depends very strongly on total utilization of the available wind and adequate purchase prices. It is noteworthy that the countries with a rapid installation rate of small WECs, - Denmark and the USA - have adopted favourable regulations for adequate pricing.

In the case of large WECs, the utilities or power generating companies will operate the units and hence control the acceptance of the machines being developed and tested. Their degree of involvement ranges from counter-publicity through scepticism to active cooperation. It can nevertheless be stated that, in countries with a promising wind energy programme, the participation of the utilities is essential.

6.2 Environmental and Safety Constraints

Although safety standards concerning the accidental fracture of WECs will be laid down in the legal installation permit, the oft-quoted potential hazards are influencing public opinion. Economy will be an automatic control mechanism for safety because, in the case of line production, heavy repair costs will hit sales. However, at many sites the formation of ice is to be expected at times, and the ice debris thrown off by the rotor may cause some danger.

Environmental restrictions stem both from objective criteria and from subjective opinions. The objective arguments deal with safety zones, interference with high-frequency electromagnetic waves for telecommunications/television, and noise. Birds have not been a problem at any installation up to now.

Safety zones are defined, e.g. near airports, and are part of the siting requirements. Electromagnetic interference occurs if metal or conducting blades are used. The well planned telecommunication systems are unlikely to be disturbed by WECs. In some rare cases of special orientation of T.V. station/large WEC/omnidirectional receiver antenna, interference with mainly the UHF channels has been reported, but good directional antennas solved the problem. Satellite communication technologies are developing rapidly, thus further reducing the already low interference potential.

In some countries, codes for permissible noise levels exist for various land uses, viz. residential areas, industrial areas, mixed areas. No regulations exist for open country outside these areas. In France, the noise-related litigation followed the "warning" of the Health Ministry (1963), that the noise emanating from a unit in constant operation should not exceed 3 dBA at night, and 5 dBA during the day, in addition to the background noise measured at the edge of the property. Noise levels of the existing WECs have been measured and shown to have such low values in most designs that restrictions based on this condition seem to be unlikely.

The other category of arguments, based on subjective opinions, mainly concerns "visual pollution". The sight of WECs is often quoted as a cause for objection. This argument has already been used in publications, hence concentrated installations of WECs should be planned very carefully with this aspect in mind.

There is also the question of preservation of nature - landscape as well as animal life - which may cause conflict with differently oriented groups.

6.3 Economic Constraints

Although about 750 small WECs (10 - 50 kW) have been installed in Denmark, the price levels, coupled with doubts as to reliability and wind availability have prevented a rapid market penetration. To overcome this barrier, financial support has been given in the form of subsidies or tax credits with the object of speeding up the introductory phase until large quantity production lowers the prices.

The recent Danish move to raise the subsidy granted for certified WECs by 30 %, has accelerated the installation rate. The American regulations which allow tax credits in various forms up to a maximum of 55 % has caused a mild boom of WEC farms in California even now, about 2 years after the publication of the Wind Energy Act.This has led to the importation of WECs from Denmark and The Netherlands.

The overall economics of wind energy are also influenced by the tariffs for repayment of the surplus energy delivered to the Electricity Boards. In the case of small WECs, favourable tariffs for autoproducers would influence their decision to install WECs and probably also induce them to consider the more cost-effective larger sizes of 50 kW or more.

6.4 Technical Constraints

Technical constraints are based on safety codes and protection requirements. In the existing units, they have been complied with in the design and installation specifications and will not represent a real obstacle in the future.

More knowledge and experience are needed in order to improve the technology and cost-effectiveness of large wind turbines. As some of the enterprises involved consider research projects to be incidental, the translation of research findings into the preparation of industrial activities does not always proceed smoothly. The transfer of knowledge and experience to the industry is a major objective of follow-up programmes. Therefore it will still be some years before reliable machines with diameters of more than 25 m can be produced in large quantities.

7. CONCLUSIONS

7.1 The Prospective Value of Wind Energy

The attraction of wind energy as a resource derived from solar energy lies in its availability as kinetic energy, hence an energy form to be converted into useful energy at a high efficiency. The main disadvantages of wind energy when compared with fossil energy are its low energy density and the lack of large-scale storage solutions; these disadvantages had a dominant influence on the results of the assessment study owing to the impact of the large structures and the necessity of combined operation in a park of mixed power stations.

The total potential of the available wind energy was found to be very high. This is true of sites suited to the installation of both small and large turbines. The limiting factor for the utilization of this energy potential was observed in the economic balance which must be respected in the combined operation of existing conventional and future wind power plants. In the order of 10 % - 20 % of the national power demands could be furnished by wind energy without major economic constraints on the existing power stations.

Furthermore, the study gives evidence that some small WECs exist in line production quality, enabling their cost-effective use in the wind-rich areas, and that the prospects for the economic installation of large WECs are also positive although the technology has not yet progressed beyond the pilot project stage.

On the whole, no negative indication has emerged which might halt the increasing interest in wind energy. On the contrary, all the findings justify the continued support of programmes for research, development and demonstration.

7.2 Recommendations for Future Research, Development and Demonstration

A large potential for more know-how and further improvements exists, wind energy technology being a young discipline. This includes technical matters as well as problems of operation and acceptability.

Many topics for additional work have been collected, and a summary is given below.

7.2.1 Wind Potential

- o develop a detailed and reliable resource atlas for the community
- o develop detailed forecasting techniques particularly to permit capacity credit for grid-connected machines,
- o set up anemological models, both for the study of wind characteristics over very large areas, as a function of the various orographic configurations, including the most complex ones, and in order to study wind behaviour in typical orographic situations (e.g., ridges, valleys, passes, etc.).
- o include the off-shore areas
- o investigate wind profiles and influences by clusters of WECs
- o develop improved gust models.

7.2.2 Siting Problems

- o develop rapid site evaluation techniques
- o investigate sites for the installation of WEC clusters
- o investigate problems of social acceptability of WECs
- o develop standard regulations for the installation of WECs, to be applicable throughout the Community.

7.2.3 Penetration Problems

- o Interconnection of clusters and equalization of energy of clusters
- o possible modifications on the conventional coal-fired plants
- o simulation studies of optimal penetration and generating mixes
- o studies on storage capacity and penetration rates
- o investigate sophisticated operational strategies for plant mixes
- o investigate integrated systems of WECs and diesel generators
- o operational experience with single WECs and clusters in grid connection.

7.2.4 Machine Technology

System Level

- o improve performance
- o develop procedures for accurate load assumptions
- o standardize design specifications
- o standardize specifications for rating and valuation of WECs
- o develop codes for European installation and operation permits
- o investigate acceptable limits of environmental impact (noise, T.V., safety, etc.)

Rotor System

- o improvements in blade technology
- o investigation of material properties for long fatigue life
- o investigate improved devices for aerodynamic control and braking
- o modify proven design for mass production
- o develop a rotor system for 100 m diameter
- o develop hardware for low M&R costs

Drive Train

- o simplification of drive-train technology
- o investigate solutions without gearboxes or reduction stages

Electrical equipment

- o develop cheap and reliable inverters
- o develop protection equipment designed to accommodate the standards needed for high penetration rates in order to avoid retrofitting
- o investigate further simplification of the generating system

Control system

- o develop devices to enable mechanical controls to be replaced by electric controls
- o investigate improved control strategies
- o study methods of advanced aerodynamic control
- o investigate joint control of WEC clusters
- o develop advanced controls for vertical axis rotors.

7.2.5 Pilot Plants

- o develop vertical axis rotors of various sizes and techniques (curved and straight blades)
- o develop simple WECs for export to, or manufacture in developing countries
- o develop small WECs for low-cost production
- o develop medium WECs for low-cost production
- o develop large WECs for low-cost production
- o development of third generation WECs
- o investigate the performance of the same WEC in different climatic sites
- o continue the study of the above item (especially penetration problems) by hardware tests
- o establish a European Test Centre, or join an existing one, to enable tests for a European type-licence.

7.2.6 Demonstration Plants

- support WEC acceptability by introduction aids
- install hybrid systems of WEC and diesel generators
- test autonomous WEC systems with storage devices
- test operational differences of dispersed versus concentrated siting
- control actual operational costs
- demonstrate the potential for new employment
- demonstrate the potential for attracting new industry with a high energy consumption in wind-rich zones.

7.2.7 Economic Aspects

- develop improved methods for the calculation of costs of a power plant mix
- investigate the overall impact of large-scale utilization of wind energy at the national economy level
- investigate the possibility of exporting wind energy.

Glossary

A	shape parameter of Weibul function: $A = a^{-1/c}$
a	parameter in Weibul function; <u>also</u>: years
ACS	average cold spell
α	exponent in Hellmann equation with respect to surface roughness
c	parameter in Weibul function
capacity credit	fraction of the investment for conventional power plants saved by the installation of wind power plants
energy credit	allowable investment for wind power plants on the basis of capitalized fuel cost savings during the lifetime of the WECs
ECU	European Currency Unit, if not otherwise stated, the exchange rates at 1 August 1981 were used: 1 ECU = 41.29 BF; 7.93 Dkr; 2.52 DM 61.60 Dr; 5.98 FF; 0.69 IRL 1250.39 LIT; 2.80 HFl; 0.55 UKL
$ECU/(kWh)_1$	specific investment for the production of 1 kWh per year
EUROPE-10	The countries of the European Communities, with abbreviations used in tables: Belgium (B), Denmark (DK), France (F), West Germany (D), Greece (GR), Ireland (IRL), Italy (I), Luxembourg (L), The Netherlands (NL), United Kingdom (UK).
8E/53N/L/4-5	identification of the subfield in the siting analysis. The example stands for the area between 8°-9° E of Greenwich, 53°-54° N, and the isovents 4-5 m/s on land.
f(v)	Weibul distribution of wind speeds
i	interest rate in %
isovents	lines of equal wind speeds
LOLP	loss of load probability
m	fuel price escalation in % per year
n	lifetime of wind energy converter in years
rated wind speed	the lowest wind speed at which rated power is produced
$\bar{v}$ mean annual wind speed	the wind speed averaged for one year
V_{10}	wind speed at 10 m height

WEC	wind energy converter
wind zones	definition in the Danish Wind Atlas is by roughness length (see Table 2-1 and Fig. 2-2).

LIST OF FIGURES

Continuation of List of Figures

Continuation of List of Figures

Continuation of List of Figures

LIST OF TABLES

Continuation of List of Tables

References

2-1 Petersen, E.L. et al. , Windatlas for Denmark, ISSN0106-2840, Risö, 1981

2-2 Frydendahl, K., Danmarks klima i vind Standard normals 1931-60, 1971

2-3 Ductiêne-Marullaz, Ph., Distributors Statistiques et cartographie des vitesses moyennes de vent en France, CSTB, Nautes, E.N. CLI 77-1

2-4 Benesch, W. et al. , Berichte des Deutschen Wetterdienstes, Nr. 147. Die Windverhältnisse in der Bundesrepublik Deutschland im Hinblick auf die Nutzung der Windkraft, ISSN 0072-4130, Offenbach 1978

2-5 Lalas, D., Analysis of Wind Power Potential in Greece, Oct. 1980 Athen

2-6 Map of Yearly Averaged Potential Wind Speeds at 10 m Height Durch Royal Met. Inst. 1982

2-7 Milborrow, D.J. et al. , The UK Offshore Windpower Resource, 4th Int. Symp. on Wind Energy Systems, Stockholm, Sep. 1982

2-8 Voith diagram 3.71-15210, Letter P40815, 7.5.82, H. Grüninger

2-9 Nielsen, P., Measurements on the Nibe Wind Turbines, Rep. No. EEV-81-04, May 1981 DEFU, Lyngby

2-10 Shaft Power Output WTS-Näsudden (AEOLUS) VFW-Report No.Ex25B-079/78, Bremen 1978

3-1 UNIPEDE? Programmes and Prospects for the Electricity Sector 1981 - 1986 and 1987 - 1990, 11th issue, Paris, Aug./Sept.1981

3-2 Janssen, A.J. et al. , Statistical Methods for the Assessment of Wind Power Integration into the Electricity Supply System, ECN-101, Petten, Oct. 1981

3-3 Former, W. and H.H. Nissen, Wirtschaftlichkeit der Stromerzeugung aus Windenergie, Energiewirtschaftliche Tagesfragen, 29. Jg. (1979) Heft 3

3-4 Jarass, L. et al. , Wind Energy, JSBN 3-540-10362-7 Springer Verlag, 1981

3-5 Martin, B. and M. Diesendorf, The Capacity Credit of Wind Power: A Numerical Model 3rd. Int. Symp. on Wind Energy Systems, Kopenhagen,1980

3-6 D'Amico et al. , Evaluation of Unconventional Sources in Mixed Hydro-Thermal System. DNU-Economic Commission for Europe Meeting, Athens, Oct. 1979

4-1 Lundsager,P., Experience with the Gedser Windmill and Small Danish Windmills, Risö Pl-ac. 2.3.1982

4-2 Madsen, B.T., Windturbines produced more than expected, Ingeniören, 23. Oct. 1981 Kopenhagen

4-3 Engineering Recommendation G 26 "The Installation and Operational Aspects of Private Generating Plant", The Electricity Council, U.K.

4-4 Engineering Recommendation G 47 Notes of Guidance for the Parallel Operation of Private Generators with Electricity Boards Low Voltage Network, The El. Council, U.K.

5-1 Jensen, P.H., En vindmölles privatökonomi, Risö-M-2335, 1982

5-2 7th Meeting of Experts, Costings for Wind Turbines, IEA, Jülich-Sper-147, March 1982, ISSN 0343-7639

5-3 Selzer, H., Wind Energy on its Way to Commercialization, Int. Coll. on Wind Energy, Brighton 1981, ISBN 0-906085-59-4

5-4 Lotker, M., U.S.-Incentives for Renewable Energy Technology, Their Use and Misuse, 4. Int. Sonnenforum, Okt. 1982, Berlin

5-5 Moynet, G., Generating Costs, Assessment Made in 1981 for Plant to be operated in 1990, UNIPEDE 10.2, Nuclear Energy Study Com., June 1982

5-6 ESB-55th Annual Report for Year ended 31st March 1982, Dublin

5-7 Jäger, F. et al. , Methoden zur betriebswirtschaftlichen Bewertung regenerativer Energiequellen, Tagungsbericht Bewertung der Wirtschaftlichkeit regenerativer Energien, Dez.1982, München

GPSR Compliance

The European Union's (EU) General Product Safety Regulation (GPSR) is a set of rules that requires consumer products to be safe and our obligations to ensure this.

If you have any concerns about our products, you can contact us on ProductSafety@springernature.com

In case Publisher is established outside the EU, the EU authorized representative is:

Springer Nature Customer Service Center GmbH
Europaplatz 3
69115 Heidelberg, Germany

Batch number: 10584448

Printed by Printforce, the Netherlands